Racecar Technology Level Three

Advanced Race Car Engineering and Applications

Bob Bolles

CRD Publishing

CRD Publications

Published by Chassis Research and Development Corporation, aka/Chassis R&D

P.O. Box 730542, Ormond Beach, FL 32173-730542. USA

Email: chassisrd@aol.com

Race Car Technology – Level Three

Copyright © 2019 by Chassis Research and Development Corporation

All photos by Bob Bolles unless otherwise noted.

All rights reserved. No part of this book/publication may be reproduced, scanned, or distributed in any printed or electronic form without permission. Please do not participate in or encourage piracy of copyrighted materials in violation of the author's rights. Purchase only authorized editions.

CRD Publications is a trademark of Chassis Research and Dev. Corp.

PRINTING HISTORY

First CRD print edition / June 2019

ISBN: 978-1-7324884-5-8

CREATED/PRINTED IN THE UNITED STATES OF AMERICA

Cover design by: LS Bulow, Graphic Designer

NOTICE: The information in this book is true and complete to the best of our knowledge. All recommendations on parts and procedures are made without any guarantees on the part of the author or the publisher. Author and publisher disclaim any and all liability incurred in connection with the use of this information. We recognize that some words, parts names, model names, and designations mentioned in this book are the property of the trademark holder and are used for identification purposes only. This is not an official publication.

Race Car Technology – Level Three
Table Of Contents – RCT Level Three

Lesson 1 - Goals of RCT Level Three	1
Lesson 2 – Concept Of Balance	7
Lesson 3 – Forces On The Race Car	13
Lesson 4 – Roll Angle Analysis & Springs	17
Lesson 5 – Weight Transfer vs. Static Loading	25
Lesson 6 – Weight Distribution Front To Rear	33
Lesson 7 – Dynamic Loading	39
Lesson 8 - Angle Of Attack Analysis	47
Lesson 9 – Advanced Front Geometry	51
Lesson 10 – Jacking Force Concept and Uses	59
Lesson 11 - Rear Steer Concepts and Uses	65
Lesson 12 - What Makes Traction?	73
Lesson 13 – Force Verses Weight	77
Lesson 14 - Where Does Force Come From	81
Lesson 15 – The Sway Bar Use As A Spring	87
Lesson 16 – Aero Downforce & Tire Loading	93
Lesson 17 - Practical Application Part One	101
Lesson 18 - Practical Application Part Two	111
Lesson 19 – Advanced Racing Shock Tuning	121
Lesson 20 – Track Tuning The Setup - Part One	129
Lesson 21 – Track Tuning The Setup - Part Two	135
Lesson 22 – Post Race Evaluation	143
Lesson 23 – Summation For RCT Level Three	149

INTRODUCTION

Around 1992 I decided to change careers and become a race car engineer. My early experience in racing as a kid was spending countless hours at Daytona International Speedway in the pits, in the stands and around the mechanics, drivers and owners listening and hearing about how the cars handled. I was always fascinated by the design and setup of race cars, be they stock cars, road racing cars, formula or weekend SCCA cars.

I attended many years of races at New Smyrna Speedway and Barberville, now known as Volusia Speedway Park. I was interested in the "race", but always fascinated by the way the cars handled and how all of that was accomplished. I watched Dick Trickle prepare his car in 1975 at a friend's garage in preparation for the Speed Weeks show at New Smyrna and thought, I would like to be able to do that.

I am an engineer by education, degree and by nature and I knew someday I would have to get involved with racing. When that day came, I threw myself into the task of learning and inventing with more energy and determination than at any point in my life with anything I had ever done. It was a passion combined with a purpose. I was determined to find the truth about chassis dynamics and race car setup.

My work, and indeed my racing business, was born out of frustration and failure in trying to find really helpful information that I could use to set up a racecar. So, I set out on a journey that followed in no one's footsteps. Instead, I used one of my greatest personal assets, a profound and acutely developed ability to apply a common sense approach to problem solving. That is exactly what you will find in this book, a common sense approach to chassis setup, vehicle dynamics and race car design, together with solid engineering theory.

This is not a controversial race car setup book and agrees in principle with technology and theory taught in major college motorsports programs. To many, according to what I hear, the books of mine that have preceded this one have become their bible of racing knowledge. Much of the technology presented here has been more recently developed over just the last five years or so. For those who believe that we had already pushed the envelope of vehicle dynamics about as far as it could go by the early 2,000's, this book does not follow that line of thinking.

Regardless of what is on the pages, the proof is on the racetrack, and the methods in this book have been tested and proven to improve performance in race cars. They increase speed, improve basic stability and have already been used to win many races and championships in many classes of racing.

How then did these RCT books come about? When I began my career working with race cars, I found plenty of information on chassis theory, but I couldn't find conclusive information that would tell me how to set up my race car in the shop the right way the first time. I had to read between the lines and keep trying different setups, working by trial and error. I personally don't like trial and error. I want to be able to know exactly how to set up my race car and know not only how something works, but why.

This book will help you avoid the trial-and-error approach to chassis setup. It will teach you sound, proven technology that is both easy to understand and easy to use, so you can set up your race car in the shop and see the positive results on the track immediately, with very little tweaking. What follows is a common-sense approach to chassis setup, vehicle dynamics and race-car design, founded on solid engineering theory. However, you will need to have an open mind, and be willing to accept new ideas that may go against previous chassis setup thinking.

Just to make it clear, the technology presented here applies to all race cars, from quarter midgets to Formula One and everything in between. This book tends to lean towards stock car racing because it represents most of the world's automobile racing. But know that not only will it be useful for all forms of circle track racing from asphalt types to dirt cars, a great deal of the technology applies to all race cars.

Success comes at all levels of endeavor, and we can't all be champions. But we can all get better at what we do. The goal of this book is to give good, solid information that has been tested and evaluated and found to be the truth. It is not, and will never be, complete as long as we continue to push the envelope in the search for better performance, but it will lay the foundation upon which future race engineers can build their programs.

Race Car Technology – Level Three
Lesson One – The Goals Of RCT Level Three

This is Race Car Technology – Level Three. It has the latest information for setting up a race car you will find anywhere. This course will cut no corners. In this course, we will present information and techniques that will vastly improve the programs for most race teams. For the others, it will provide performance you never thought possible, even for current championship winning teams.

The goals for designing our race cars are basically the same be it a formula road racing car or a late model circle track car. Performance gains in the slower portions of the track cause all of the speeds to increase around the track including on the straights.

A True Statement: 99% of the crew chiefs, race engineers, or whomever, who are in charge of setting up a race car do not know how, and do not have the tools, to setup a race car without trial and error. 15% of these are decent at tuning their race car setup to the point of winning. The other 85% are just along for the ride.

The proof is this. How many teams at the local level win during the season, 2, 3, 4? Out of 15 to 25 entries? That is anywhere from 8 to 20% +/- of the teams that show up that can win and even then, they don't win all of the time.

What about in the most expensive form of motor racing where hundreds of millions of dollars are spent each year and where the most educated engineers and technical chiefs reside? On a good race day, only three Formula One cars out of twenty-four are competitive and in a position to win for a 12.5% performance rate. Most of the time it is two cars for an 8% performance rate.

Explain to me how, with all of those resources, can the bottom 25% of the cars be a full two to three seconds slower than the leader on the same tires? A tenth of a second for an average F1 race track is about 15 feet in one lap. If you are one tenth faster than another car, you will gain 15 feet per lap. Two seconds is 20 tenths. 20 times 15 is a whopping 300 feet PER LAP that the fastest car gains, or moves ahead of, the back 25% of the cars, on the same tires.

If you follow F1 and watch closely, those same bottom 25% cars actually have very quick straightaway speeds. The gains that the leaders have are through the slower turn portions of the race track where aero downforce is lowest because the speeds, something aero needs, are lacking. A great deal of the gains are with enhanced mechanical grip and few of these teams have maximized mechanical grip.

If all of that educational power and knowledge gained through higher education was correct and perfect throughout that industry, we wouldn't see this much difference in performance. Obviously, most of these engineers lack the knowledge and the tools to get the job done.

Another True Statement: Almost every team that wins does not get the most performance out of the car. How do I know this? By experience. Just because a team wins does not mean that car is as fast as it could be, it's just faster than the competition, in that race, at that race track.

Major track championships and major series championships have been won by setups that are not the best they can be. This is because the person in charge of the setups didn't have the knowledge and/or the tools to be able to setup the race car to be the best it could be.

It's not the desire of the crew chief or engineer, nor the effort that keeps them from setting up the race car as good as it can be, it is the lack of knowledge and tools that holds most crew chiefs back. The education for motorsports engineering must be lacking in today's racing world.

In this course, we will give you the knowledge and the tools to get the job done.

What We Will Learn In Level Three - At this RCT Level Three point in time, we would hope that you have completed Levels One and Two. In those Lessons, we learned about all of the parts and pieces of a race car, and we also learned how to adjust and position all of those parts so that they can work together and not hinder the actual race day setup. So, now we come to the actual setup part and how the winning teams perform their magic. We will setup actual race cars.

I say it is magic, but it is not really magic at all. It is knowing what to do, in what order to give the car what it wants, and that is a setup that takes advantage of aerodynamics, mechanical balance and tire contact patch optimization.

When we get the car as low as possible giving it a low center of gravity and better aero, and when we arrange the spring rates and moment centers so that the load distribution on the four tires is ideal, and when we create the optimum geometry so that the tire contact patch is as large as possible, we will then have a winning race car. All the driver has to do after that is drive it to victory.

In this Level Three Course, we will show you how to do all of the above, exactly like every winning team does it, or at least how they should. Even winning teams get some of this wrong. But how do they win, you might say? They win because everyone else they are racing against isn't 100% right either.

The winning team is often at 90-95% of the perfect setup, when everyone else is 85-90%. So, there is room for improvement in your race cars setup and beat those dominant teams that are not all that they can be.

Never think that you cannot improve what you are doing. Even if your car is perfect with the current level of knowledge, there may be something out there that you can learn and that can take your division to a whole new level. Then you will be top dog, at least until everyone figures out what you are doing.

One critical component in doing well in racing and being successful is consistency. What we will strive for in this school is impressing on you the fact that you have to keep everything under control and in optimum condition. We can never let any of the settings or routines get out of hand. Many winning teams run the very same setup every race of the year, even at different, but similar race tracks.

We test within limits and we always go back to what we know works. This is called the baseline and this school takes you to the baseline. And that baseline is often what wins races and continues to win races. We can step out of that baseline to try and find more performance, but 9 times out of 10 we won't improve and that is why we go back to the baseline.

We often think of the formula one cars as being the ultimate race cars and perfect in every way. If that were so, then there wouldn't be a two second gap between the fastest and slowest cars. There must be a difference in designs that causes that huge gap in speeds. We like to think it is lack of mechanical grip that makes not only the slow turns faster, but the entire lap.

What We Should Have Already Learned – RCT Level Two taught us how to prepare the car for this Level Three Course. If you have not taken Level Two, then you might miss something important that will interfere with what we are about to do in Level Three. It happens all of the time, and you are no exception. Here is what we said about Level Two in that courses introduction.

Race cars don't just come ready to race. You might be buying a brand-new race car, or starting to work with a used race car you just acquired, or maybe you are building your own race car. In any event, you must go through the car and setup each system of that car properly and in a specific order.

No one gets away with not doing any of this. You cannot be successful in racing if you don't work with these setup routines. And even if the race car manufacturer or previous owner says it is good to go, you cannot trust anyone but yourselves. We must verify each and every setup parameter of the race car.

If you are racing now, or have watched others race and have asked how they do it to make their cars so successful, it isn't magic and no one ever falls on a winning setup by accident. No matter how easy it all looks, a lot of effort and smart thinking goes into every winning race car.

In the Level Two Course, we performed simple, but necessary functions like determining ride heights. We set the weight distribution, and establish proper caster and camber for your application. We went through all of the various alignment settings and carefully explained why we are concerned with each phase and how those will make the car perform.

We established what spring rates to use and which sway bar will match those spring rates. We matched our shock rates to the spring and sway bar rates too. We chose tires and tire sizes and learned how to select tires and prepare them for competition. Then we put this package to the test and fine tuned it at the race track.

When you got done with the Level Two Course, you were able to take a modern day race car and set it up. When I first started into racing, I could find no good information to tell me how to setup a race car. Even today, information comes at us in parts and pieces and it is very difficult to put it all together into a winning combination. The Level Two school was designed to provide all of the necessary information to guide you through the setup of your race car.

The Ultimate Goals - Our ultimate goals are as explained in the RCT Level One Introduction. For those of you who forgot, or those who chose to begin your schooling with RCT Level Two or Three, we urge you to go back and take the Level One and Level Two courses. In our introduction to Level One, we explained what our overall goals are for designing a high performance, winning race car. It bears repeating, so we will again explain.

The race engineer's goals are very simple in concept, but more complicated to carry out. On any race course, be it with circle track or road racing, on a dirt or asphalt surface, there are key areas of performance where your car needs to be made better than any of your competition and those are the following;

Maximizing Grip - When non-chassis elements like powerplants are more or less equal, Grip is what will help you win races and championships. Motors accelerate you, brakes slow you, but Grip makes you faster through the slower turn portions of the track and relatively small gains in lateral Grip will produce huge gains in speed and performance.

What are the Key elements of Grip? - The tire contact patch is where Grip is produced. Causing the tire contact patch to: 1) be larger, and 2) causing the most vertical loading on that tire contact patch are the two key ingredients for maximizing Grip.

Maximizing the amount of traction that is available from the four tires on a race car, any race car, will make you as fast as you can be, all other things being equal. Everything we present in RCT L3 will ultimately lead to optimization of the race cars Grip and the use of that Grip to go faster.

What are the basic parts that make up Grip?

● Loading On The Tire – The more load we can put on a tire, the more Grip that tire will have, period. But, the gain in grip is not linear as we will explain later on. Load can come from the weight of the car, mechanical downforce from banking, and aerodynamic downforce, all of which will be explained in detail later on.

● Contact Patch Area – The greater the size of the contact patch, the more Grip. If we can find ways to make the tire contact patch larger, then that tire will produce more Grip.

● Tire Compound – The physical and chemical makeup of the tire can provide more Grip. The softer the material, the more Grip we will have within limits. We have rules that govern the softness of the tires, but we need to stay very close to those limits.

● Load Distribution – A pair of tires on the same end of the car, or same axle, will produce more Grip when they are more equally loaded. The most Grip from a pair of opposing tires will come when they are equally loaded. There is a variation to this concept for dirt racing that will be addressed later on.

● Angle of Attack – What is called Angle of Attack, or Slip Angle, is when a tire is pointed slightly to the inside of the arc it is following through a turn. If a tire were to follow the exact tangent line around a curve or arc of the turn, it would not generate any side force to counter the centrifugal force.

So, in consideration of the other items that make up Grip, it is fair to say that none of those would be useful if it weren't for the creation of Angle of Attack. No matter what amount of Loading or what the size of the Contact Patch area is, or how soft the tire compound is, or how equal the load distribution between opposing tires, the car would not stay on the course without the tires developing an Angle of Attack.

So, there you have it. Those five things represent the parts that help make the Grip we seek to make us faster so we can win races. As we go through each part of the race car in the other Lessons, we will explaind how to optimize those parts to enhance our Grip and make us faster through the slow speed turns. And we will understand how that will in-turn make our whole lap faster at every point around the course.

Why Does More Grip Make Us Faster?

When a race car turns, a lateral force called Centrifugal Force tries to push the car to the outside of the turn. The tire contact patches resist this force. The speed we can drive through the turn is limited by the amount of Grip we have in our tires. The more Grip, the faster we can drive through the turns.

The one often overlooked benefit of achieving faster turn speeds is this: the faster you exit a slow speed turn, the higher the speed at which you will start accelerating down the straight part of the track in-between the turns. So, speed gained in the slow speed turns will be carried down the straights too. It's not just that we gain speed in the slow speed turns with more Grip, we gain everywhere around a race track.

To give you an example, on a typical or average length Formula One track, one tenth of a second is about 15 feet on the race track. Some teams are a full second slower per lap than the fastest teams. That is 150 feet per lap that the faster cars move ahead of the slower cars each lap. In a fifty lap race, that equals 1.4 miles, or 2.3 kilometers.

We already know that some teams get lapped in a F1 race, and the average length of those tracks is around 3 miles. So, those lapped cars are on average 2.0 seconds per lap slower than the winning team. And they all run with the same choice of tires. That is hard to fathom.

Is the gain realized by the winning F1 teams all Grip? No, as we have read, some engine packages are down on horsepower and that is a factor in the difference in lap times. But what about the teams who have the same engine package as the winners? Why are they so slow? It could be that they lack mechanical grip in the slower portions of the race track.

Building mechanical grip is one of the most important areas of race car engineering, even over and above aerodynamic grip. The reason is this; To gain maximum aero grip, you must have high speeds and the most gains from mechanical grip happen in the slower turns where the speeds and the aero downforce is lower. So, the only thing we can point to as the reason for the gain in performance for the winning teams is Mechanical Grip.

The Concept Of Mechanical Balance - If your car has more overall Grip than other cars early in the race, that Grip might not stay superior and you might end up being a slower car through the turns later in the race. This is due to the Balance Factor. There are two definitions of balance. They are not the same and they cannot be confused.

One is handling balance which is defined as a car that is neither pushing (understeering) or loose (oversteering). Pushing is when the front set of tires has less Grip than the rear set of tires. Loose is the opposite.

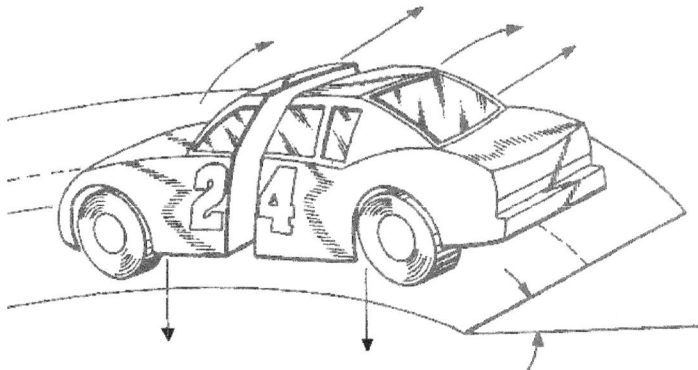

The Mechanical Balance concept is simple. We want both ends of the car to work in sync so that one does not try to cause the other to do what it does not want to do. When the two ends of the car are working in unison, we can achieve the goal of mechanical balance.

A team can attain handling balance fairly easily, but just the fact that the car is neutral in handling doesn't translate to speed or consistency. A car can have less overall Grip than other cars and still be neutral in handling. Although a neutral handling car is a goal, it is not the first and primary goal and not what we are discussing here.

The other balance is Mechanical Balance. Simply put, this is when the two suspension systems, front and rear, are working in sync and where the load transfer is predictable and maximized. This is what teams need to search for. The end result of all we do with chassis engineering must be towards the goal of Mechanical Balance.

Circle track race cars are complicated, but much easier to setup that a road racing car. This is because the setup in a circle track car can be asymmetrical, meaning the spring rates, cambers, etc. from side to side can be different. The circle track car is only turning in one direction and so the setup need only be correct for that direction of turning.

The Search and Desire for Mechanical Balance should be the foremost goal for every race engineer. A lot of good things come with achieving MB. The tires do not work as hard and will last longer. The car is much easier to drive. The car maintains higher turn speeds as the tires wear later in the race, or later in each stint for cars that are allowed to pit and take on new tires.

When you have finished this Level Three Course, you will have all of the tools necessary to complete the setup of your race car. All that is then needed is fine tuning, which we will also cover. If you do it right, your car will have the Mechanical Balance needed to run fast for a long period of time. Your tires will wear more evenly and at a slower rate than if you were not balanced. And you will have the speed at the end of the race that will carry you to victory. So, let's get started.

Race Car Technology – Level Three
Lesson Two – Concepts Of Balance – In Depth

The concept of balance in a dynamic sense must be understood before we can get into any of the other areas of discussion in RCT Level Three. This theme represents the basis of everything we do from this point on. The concept has probably been around for a long time, but in the mid-1990's, a method was invented to improve the way we measure the balance in a race car's suspensions.

Previous attempts to create a measure of balance involved measuring the roll stiffness in each suspension system, front and rear. What we will discuss in this and future Lessons is not that. It is similar, but more refined and takes into account effects that act in addition to the roll stiffness and what has been referred to as roll couple.

The Basic Premise – The basic premise is this. The two suspension systems in a race car, any race car, will have spring rates, spring split and roll / moment center heights that will determine how that suspension system reacts to the lateral forces we encounter during cornering.

The "React" we are talking about in this example is made up of two things:

1) The chassis will, in most cases, dive or compress, especially if the banking angle of the turn part of the track is significant. That is, the center of the chassis will be lower through the turns than is was at normal ride height. If both the front and rear suspension systems dive the same amount, or differently, there is no negative influences associated with the dynamics of the setup.

2) What each suspension system also does as a reaction to the lateral forces that exist when going through the turns, is to roll. Every suspension system will want to roll to its own individual angle that can be calculated. And each pair of suspension systems would do just that except for the fact that we have a strong and resistant chassis connecting the two suspension systems.

If the two suspension systems want to roll to the same angle, then the overall setup in the race car is what we call balanced. If the two roll angles each suspension system want to roll to are different, then we call that an un-balanced system. And that is not to say that we might, in certain situations, want to purposely cause unequal roll angles for reasons we will get into later on.

For now, a balanced setup is one that has equal roll angles.

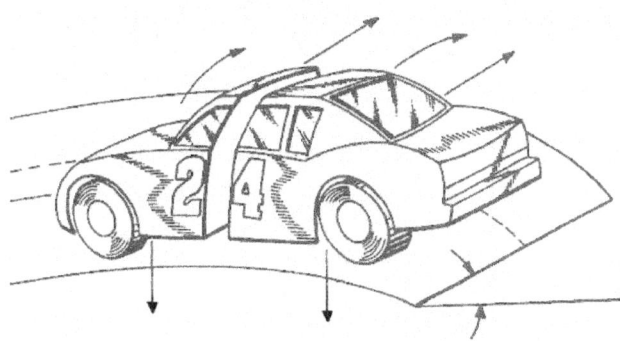

The concept of Roll Angles is how we define balance. Each of the suspension systems at each end of the race car will seek their own roll angle based on the spring rates, moment center heights, etc. The Roll Angle concept takes into account all of the influences that cause the car to want to roll to its desired angle. Our goal is to match these two roll angles.

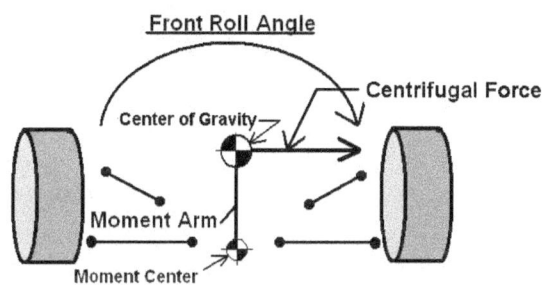

The roll angle in a AA-arm suspension is a product of the length of the moment arm, the amount of centrifugal force and the spring rates translated out to the wheels along with the track width. We want to match this roll angle to the roll angle at the other end of the car. When we do that, the distribution of the weight that transfers becomes very predictable and the handling becomes very consistent.

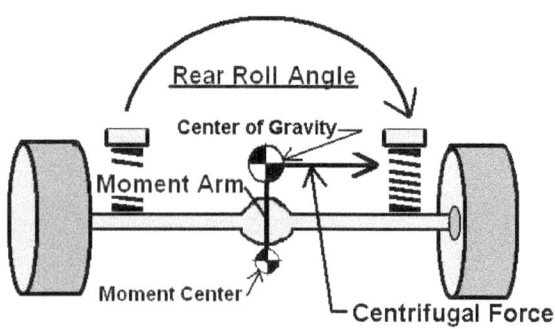

In a solid axle suspension system, we also see a roll angle that we need to match to the other suspension system. This angle is a product of the length of the moment arm, the amount of the lateral force and the spring base width, which acts much like the track width at the front.

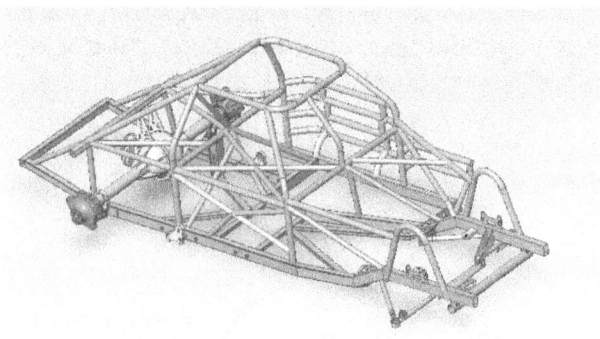

The reason why we need to match the desired roll angles at the front and rear is because each of these suspension systems cannot roll to their own angle when the other suspension system is wanting to roll to a different angle. We have a stiff chassis connecting the two systems that prevents each one from being independent. That is why we need to match the desires of each.

Why Does It All Come Down To Roll Angle? – The reason why we reduce all of the influence of lateral force and banking angle down to the roll angle is because the roll angle is the end result of all of the influences. Roll stiffness does not take into account some very significant influences. And that is why we cannot use that methodology to attempt to measure the balance in race car suspension systems.

The inventor of that method was trying to find a way to give the race car what it wanted for setup. The overall setup involving the front and rear suspension systems was considered and it was concluded that if we can cause the front and rear suspension systems to work together and not fight each other, then that is what the car wants. And this concept proved to be true.

In subsequent testing, it was found that the car reacted very well to the balancing of the roll angles. The first thing good that happened was that the weight transfer became predictable. That is, we could now calculate the weight transfer at each end of the car to know how to set the static weights so that at mid-turn, there would be ideal loading on the four tires. This would produce the most traction possible.

When real race cars were analyzed and the setups configured so that the roll angles were equal, in a stock car racing on a circle track, or on a road racing application, the car became faster, the speed stayed with the car longer, and the tire wear was much better.

So, now that we have a better understanding of what balance is, let's examine what influences the roll angles and in what way.

What Influences The Roll Angles? – The following items influence the roll angles in a race car.

1) The *Spring Stiffness* in the suspension system. The stiffer the overall spring rates, the less roll angle we have. This is common sense, because a race car has very little roll with its very stiffly sprung suspension system while a family sedan wants to roll quite a bit. One has a very stiff suspension and the other a very soft suspension.

2) *Spring Split* – Because of the way the forces work in a race car, using different spring rates on each side of a suspension system will influence the roll angles. A softer spring on the outside will increase the roll angle.

3) *Spring Base* – The width of the spring base does influence the roll angle. The wider the base, the less roll angle. For a straight axle suspension, the spring base is the width between the top of the springs, or coil-over top mounts. For a AA-arm suspension, the spring base is the track width, or distance between the centers of the tires.

4) *Magnitude Of The Lateral Force* – The greater the lateral force, or what we call the G-force, the more roll angle we will have. This makes sense too because as we corner slowly, the measured roll angle is less than when we go faster through the turns. This is why semi-tractor trailers tip over when going too fast on the off-ramp. The roll angle exceeds the suspension travel.

5) *The Location Of The Center Of Gravity* – What acts to roll the car is the overturning moment, something we'll get into in the next Lesson. The CG is at the top of the overturning moment arm and the higher the CG is, the longer the moment arm is which causes more roll angle. The location laterally also has an influence on the roll angle. The farther left the CG is located, the less the roll angle will be.

6) *The Moment Center Height* – The moment center height is the bottom of the overturning moment arm. As such, the higher the MC, the shorter the moment arm is and the less the chassis will want to roll.

7) *Sway Bar / Anti-roll Bar* – The sway bar, also called an anti-roll bar influences the roll angle. It can be changed to influence the roll angle for cars equipped with those devices. Either the diameter of the bar can be changed, or the arm length can be changed to influence the roll angle. The larger the bar diameter, the stiffer the bar and the less roll angle. The longer the arm, the softer the rate of the bar and the more roll angle the car will have. The reverse is true for each of these changes.

8) *The Track Banking Angle* – The banking angle of the turns from horizontal, does influence the roll angles. Imagine the car racing around a track where the banking is 90 degrees from horizontal. The car would only want to compress into the track and not roll at all. There would be no overturning moment because the lateral forces would be acting straight down from the cars perspective and not to the side. The track banking angles we usually see have a similar effect and act to reduce the roll angle. We'll get a better understanding of that in the next Lesson.

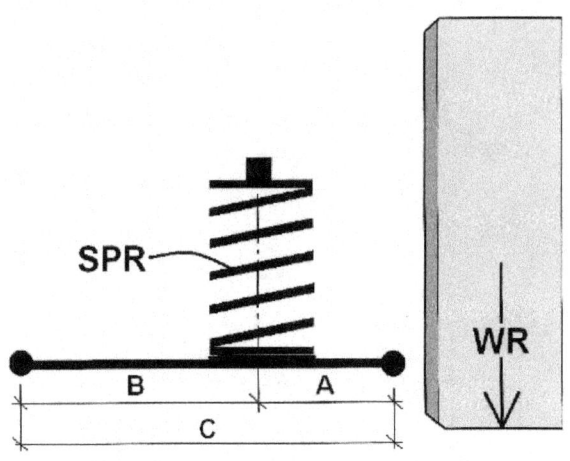

The same motion ratio concept is true of this big spring, or stock spring, design. There is a motion ratio that translates the installed spring rate out to the wheel to become a wheel rate we use in determining the roll angle.

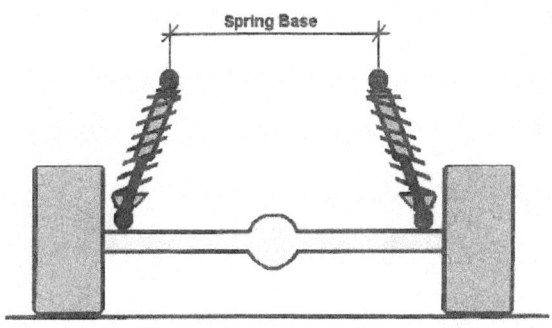

In a solid axle suspension, the spring base is the distance between the top coil-over mounts for this coil-over installation. The chassis rests on the top of the springs and this is all the suspension knows for spring rate and spring base.

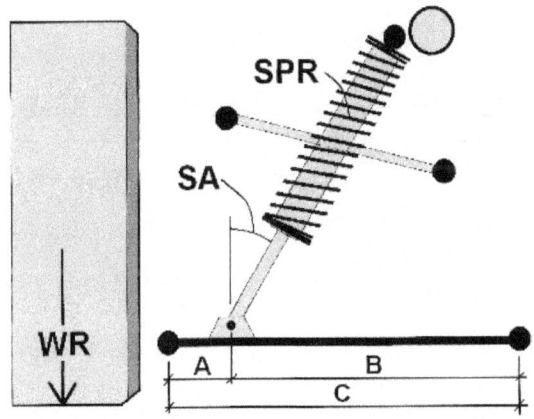

The installed spring rate in a AA-arm suspension is translated out to the wheel through a motion ratio. Here we see a coil-over installation. The track width (distance between the centers of the two tires) helps determine the roll angle. The wider the width, the less the roll angle. Then the wheel rate becomes the spring rate for a AA-arm suspension for the purpose of determining the roll angle.

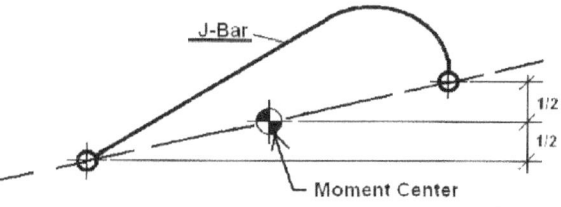

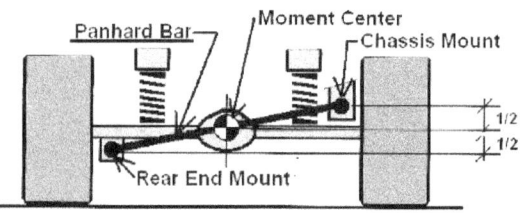

In the rear of a stock car, or any solid axle suspension, the moment center is also the bottom of the moment arm. When we move the panhard bar, or other device used for lateral location of the rear end, up or down, we are shortening or lengthening the moment arm. This has an effect on the amount of the roll angle. A longer moment arm yields a larger roll angle.

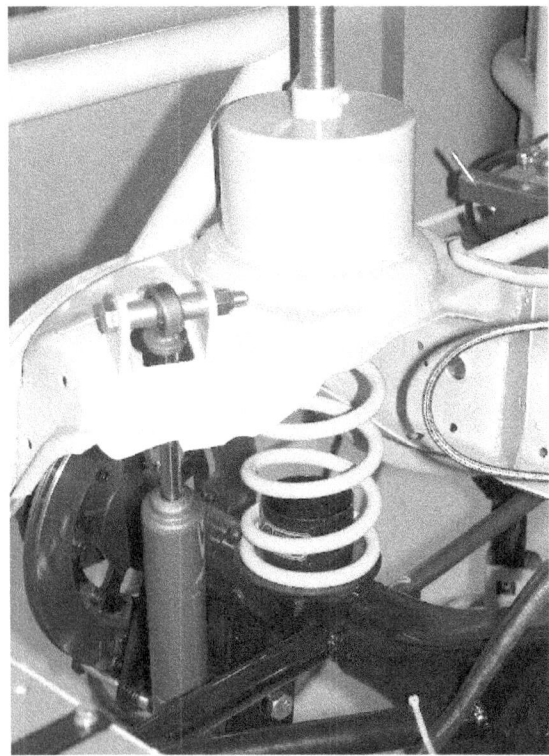

In this stock rear suspension, the springs are mounted straight up, unlike the coil-over springs on a typical asphalt or dirt late model. The distance between the centers of the two springs is the spring base. The wider the spring base, the smaller the roll angle.

Because spring split, or having a spring on one side of the car stiffer or softer than the spring on the other side, is so influential, we need to know the exact spring rate for every spring we install in our race cars. If we install a spring rubber, or bump stop, we then need to rate that spring in the range of motion it will be moving within on the race track so we can know how to arrange the spring rates.

What Can We Change To Make The Car More Balanced? – Out of the above influences, there are some that we can easily change to cause the roll angles to change. These are:

1) The *Spring Rates*,

2) The *Spring Split*.

3) The *Height of the Moment Centers* for both AA-arm and solid axle suspensions,

4) The *Spring Base* – In some designs, especially for dirt cars, the rear spring base can be easily adjusted by moving the tops of the coil-overs in or out to change the spring base width.

5) *Sway Bar* – Teams can change the sway bar settings, or size of the actual bar, to increase or decrease the roll angle.

All of these are easily changed and those are the changes we mostly use to setup our race cars.

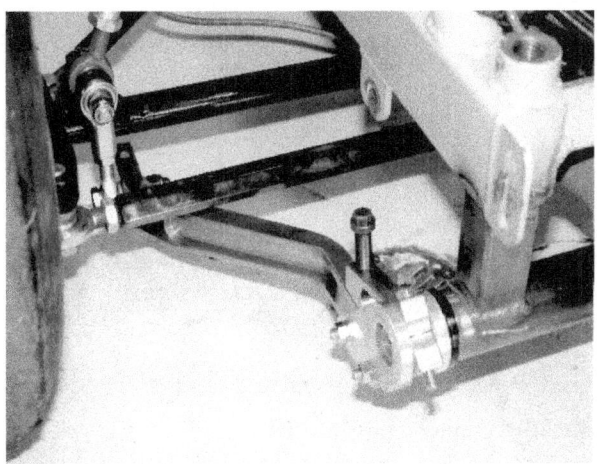

The sway bar influences the roll angle. There are many different types and designs of anti-roll systems. This one is used on an asphalt late model car and is adjustable for pre-load. The sway bar becomes a spring much like the ride spring, when it is made to twist by the dive and roll of the chassis. A larger, or stiffer, bar will result in a smaller roll angle.

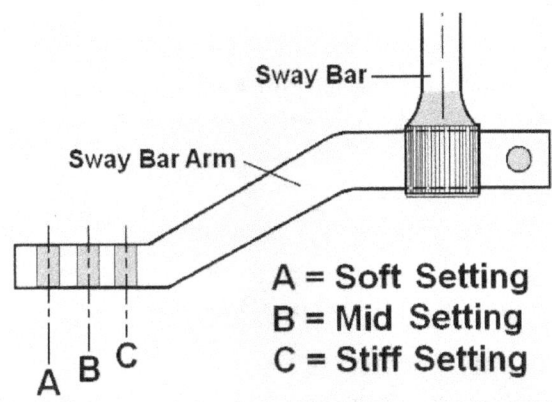

The length of the sway bar arm helps determine the usable spring rate of the bar. In the A position, the arm is the longest and the rate is the softest. In the C position, it is opposite, it becomes the stiffest setting.

In this example, the sway bar system consists of a bar and a set of two blades that act as the sway bar arms. These blades can be rotated so the when flat to the direction of pull they will bend and soften the overall spring rate of the sway bar system. When turned to be long ways to the direction of pull, they do not bend and create the stiffest setting for the sway bar system.

How To Use The Items To Change The Roll Angles – Here are some ways, in general, that we can change the roll angles in the suspension systems. It is good to know, at this point in the RCT Level Three, what can be done to change the roll angles and with those, the balance of the setup.

Spring Rates – We can increase the overall spring rates to reduce the roll angles. The stiffer the spring rates, the less roll angle we will have. And conversely, the softer the spring rates, the more roll angle we will see. So, if we need to reduce the rear roll angle to match the front, we can stiffen the rear spring rates as an example.

Spring Split – We can also increase or decrease the roll angle by changing the difference in spring rates on each side of the car. This is usually only done for circle track race cars. For example, we could soften the left rear spring rate and/or stiffen the right rear spring rate and that would reduce the rear roll angle. This works the same way at the front.

Conversely, if we want to increase the rear roll angle, we can increase the left rear spring rate and/or decrease the right rear spring rate to increase the rear roll angle. This works the same way at the front.

Spring Base – In certain race cars, the top spring mounts can be moved closer or further away from the centerline, mostly on straight axle suspensions, to change the roll angle. Moving them in towards the centerline increases the roll angle by shortening the spring base width. Moving them out away from the centerline reduces the roll angle.

Moment Center Height – We can change the moment center height at each end of the car to change the roll angles. The changes to the AA-arm suspensions will be more difficult, but not impossible. Changes to a panhard bar / straight axle system will be much easier in most cases.

Sway Bar – To tune the balance of the car, we can make adjustments to the sway bar. At the track, we will usually change the actual bar in a stock car running either circle tracks or road racing circuits. Less common changes include changing the length of the arm attached to the sway bar. The shorter the arm, the stiffer the bar rate will be and the less roll angle we will see.

For road racing prototypes and formula applications, there is usually a system that includes a sway bar and arms that are bendable blades with different thicknesses. The designs have the ability to rotate the blade so that they bend more or less easily, which stiffens or softens the bar. If the balance, and roll angles, are way different, then a sway bar diameter change is required.

Summary - This Lesson is relatively short, but very important. The very concept of balance and suspension roll angles is at the very heart of everything we do with the springs, moment centers, etc. If you can grasp this concept and use it to plan out your setup, then you will be way ahead as a race car crew chief, designer and engineer. From here it is a natural progression to talk about the Forces on a race car in the next Lesson Three.

Exam - In The Context Of This Lesson:

The Most Efficient Method Of Measuring Balance In A Race Car Is?

1) By determining the roll stiffness
2) By determining the roll angles
3) When we can measure the shock travels
4) Related to spring stiffness

Roll Angles Are Determined Using Which?

1) Spring rates
2) Moment Center heights
3) Spring base
4) Spring split
5) All of the above

The Center Of Gravity Is The?

1) Top of the moment arm
2) The bottom of the moment arm
3) Always at the center of the car
4) Different for the front and rear of the car

The Front And Rear Suspensions Cannot Roll Freely Because?

1) The spring rates are different
2) The moment arm heights are different
3) They are connected by a stiff chassis
4) The forces at the front are different from the rear

The Spring Base For A AA-arm Suspension Is The Track Width?

1) True
2) False

The Spring Base For A Solid Axle Suspension Is The Width Of The Top Of The Springs?

1) True
2) False

The Sway Bar/Anti-roll Bar Does What?

1) Assists the roll angle
2) Resists the roll angle
3) Acts like a ride spring in roll
4) 2 and 3

A Longer Sway Bar Arm Does What?

1) Makes the sway bar stiffer
2) Makes the sway bar softer
3) Increases the spring stiffness in roll
4) Increases roll stiffness when a smaller bar is used

Race Car Technology – Level Three
Lesson Three – The Forces Acting On The Race Car

In this lesson we will learn about and study the forces that act on a race car. We won't make this overly complicated because we want you to both grasp the concepts and then be able to think through how they affect the race car. This will lay the foundation for much of what follows in the Level Three course.

To be complete when talking about forces on a race car, we have induced forces from two different directions. There is the Lateral Force, which acts on the car when we are cornering and the car is changing direction. The other force is Longitudinal Force that acts on the car when the car is accelerating and when it is decelerating.

We want to now concentrate on only the Lateral Force and we will take up Longitudinal Force later on. Trust me, Longitudinal force is important, but it applies to other areas of setup than what we are getting into within the next few Lessons.

There are some very complicated and sophisticated dynamic and kinetic measurement rigs available to the race teams. These, like this example at Morse Measurements in North Carolina, will try to duplicate the forces we record acting on our race car when we are cornering, braking and accelerating.

What Are The Forces We Need To Know About? – The obvious force we can feel when the car corners is the Lateral Force. It tries to roll the car and it tries to push the driver to the outside of the turn. This force is rated at G-forces. The G-force is the sprung weight of the car in multiples directly related to turn speed. The faster we go through the turns, the greater the G-force. The number, 2.0 G's represents the sprung weight times two.

There is a formula for calculating the G-force. It calculates the total lateral force and then divides that by the sprung weight. It is:

G-force = [(Sprung Weight x Speed in MPH²)/(14.97 x Turn Radius in feet)] / Sprung Weight.

So, if we had a race car that weighs 2100 pounds in sprung weight, and was going 80MPH through a turn having a radius of 200 feet, the G-force would be the total lateral force (the part in the [ ]) divided by the sprung weight, or 2.14 G's.

The formula cars show us a lot about lateral forces and how cornering produces this force. Many F1 cars can reach or exceed 4 G's in the turns with the aid of Aero Downforce. This represents four times the sprung weight of the race car when we are using lateral G's for the calculation of forces and how they affect the roll and weight distribution of the car.

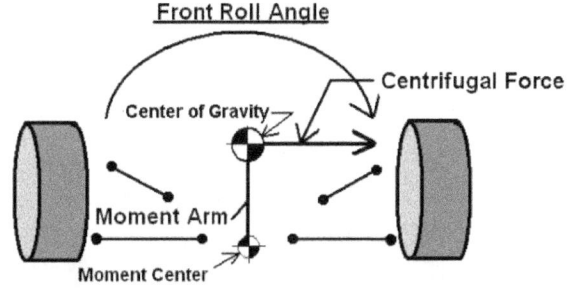

A very simplistic way to look at cornering forces is to just consider the lateral force, or what we call the Centrifugal force. This is what is trying to force us off the turns and what the tires resist. We cannot, in the study of the forces on a race car, just think about only the lateral force. There is more to the equation, so to speak.

13

There is another force acting on the race car when it is cornering, and everywhere else it is situated. That force is Gravity. Gravity acts on the race car when it is at rest as well as when it is cornering. It never goes away. When we analyze the forces acting on the race car, we have to consider both the lateral force and the gravitational force.

In the study of physics, there is a term called Resultant Force Vector. This Resultant vector is a magnitude and direction of a force that represents a combination of two forces. That is why it is called Resultant, it is a result of the existence of two forces. In our race car we do have two forces, lateral force and gravitational force and they both act at the center of gravity of the sprung mass. So, we too have a Resultant Force Vector.

The concept and existence of a resultant force vector helps us understand a lot of what goes on with a race car during cornering. And we will get into all of that. But first, I don't want to lose you on this concept and how it works. It does sound complicated from a distance, but once we get closer to it, everything will make perfect sense and be easy to understand, I promise.

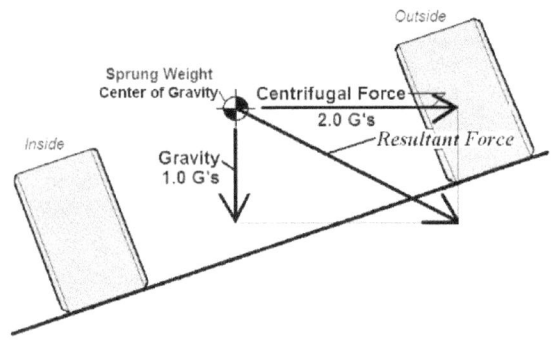

This is a more accurate and complete way to look at the forces acting on our race car and the suspension. There are two forces that act on the sprung mass of the car, and the both act at the sprung mass center of gravity. The first and ever-present force is Gravity. It is always there pulling the car towards the center of the earth, even when we are racing and cornering. The other is the Centrifugal force that is pulling on the car parallel to the earths surface. Its magnitude is determined by the weight of the sprung mass of the car, the speed, and the turn radius. Those create a force and that force divided by the sprung weight gives us the G-force number.

What Does A Force Vector Look Like? – When we have a force vector resulting from two forces, it will look like the sketch shown above. The length of the two force lines are used to form the resultant force length. The length of the gravity force line is then multiplied times the G-force number, such as 2.0, to determine the length of the G-force line.

From the ends of these two lines, we draw lines that are parallel to the opposite line to form a rectangle, or in some cases with force vectors, a parallelogram, meaning all sides are parallel to the opposite side. In our case, with a race car, the object is a rectangle because the gravity line is always at ninety degrees to the lateral force for almost every turn condition at almost every race track. Or, gravity is a force pulling towards the center of the earth and the lateral force is pulling parallel to the earths surface.

How Does The Force Vector Concept Help Us Understand Our Race Car? – Many people who try to understand vehicle dynamics forget about gravity. They only concentrate on the lateral force. The direction and the magnitude of the force acting on the race car matters. If we think of the dynamics in terms of a force going in a much different direction, any conclusion we come to has to be wrong.

Here is an example of why it matters. Suppose we have a typical race car going through a flat turn and gravity was not acting on that car. Only the lateral force is pulling on the car. The car would roll such that weight would come off the inside (inside of the turn) springs and be put onto the outside springs.

In this scenario, the inside of the car would raise up and the outside of the car would compress. But, our observation of most race cars shows us that this is not what happens. For most race cars, the outside of the car compresses, but the inside does not raise up and in some cases with high banked tracks, will compress too.

If both sides are compressing, then some force must be pulling down on the car. Note in illustration RCT-C3-L3-YY that the resultant force vector is pointing down and to the outside and meets the ground at a point between the two tires. Won't this direction of the force, and the fact that it is directed between the tires mean it will be doing two things, one rolling the chassis and the other, pulling down on the chassis.

The higher the banking angle, the closer to the centerline of the chassis the resultant force will move. That is why on very high banked tracks like at Daytona, the cars that ran before coil bind and bump setups rolled very little and compressed a lot. For most medium banked race tracks, the outside compresses and the inside of the car stays relatively at the same height as it is at normal ride height.

On a really high banked track like Bristol Motor Speedway, the direction of the Resultant Force is more towards the center of the car and we then have less roll angle and more mechanical downforce. The G-force is much higher than we see at flatter tracks. This force pulls down on the car more than it tries to roll it over. The setups for this type of track are much different than those for flatter tracks. The springs must have more rate, much more, and the balance is different, all due to the high banking angle.

The Effect Of Track Banking Angle – The banking angle of the turns has three effects on the performance of the car. First 1) the high angle changes the location of where the resultant force intersects the ground keeping more load on the inside tires creating more grip, 2) as a result of the vector angle, there is more mechanical loading on all of the tires causing the car to be able to go faster, which is why the G-force number is so much higher, and 3) the resultant force is higher due to the increased speed creating even more tire loading through the turns on high banked tracks. A faster speed generates the greater G-force number. Note that in the formula for calculating G-force, speed is a determining factor in the calculation.

Since the force is directed more downward in relation to the track surface, and since it is greater, there is more loading on the tires than we would see on flatter tracks. This provided more grip helping the car to go faster. On tracks like this, the turn speeds can increase by almost double over a flat track.

Summary For Lesson Three – Please go back and read this over again until you can fully understand this concept. The way these forces act on the race car chassis determines everything we do to enhance performance. As we race at different tracks with different turn radii and different banking angles, we need to understand how these forces will change in magnitude and direction. We must then make changes to our setup in order to meet our performance goals.

Exam - In The Context Of This Lesson:

The Two Forces Caused By The Turning, Braking and Acceleration Are?

1) Lateral and Longitudinal forces
2) The braking force
3) The accelerating force
4) The Earths rotational force

The Two Forces That Act On The Race Car In The Turns Used For Handling Are?

1) Lateral and Longitudinal forces
2) The braking force
3) The accelerating force
4) Centrifugal and Gravity

The Resultant Force Direction Is Determined By?

1) The magnitude of the Lateral force
2) The weight of the sprung mass
3) The front and rear moment centers
4) 1 and 2

Which Of The Following Are Used To Determine The G'force?

1) Speed of the car
2) Sprung weight of the car
3) The radius of the turns
4) All of the above

High Banked Tracks Cause <u>What</u>?

1) More mechanical downforce
2) Higher speeds
3) Less roll angles
4) All of the above

Race Car Technology – Level Three
Lesson Four – Setup Platform Analysis & Springs

This Lesson is about the setup Platform, as many call it. We can also call it the synchronization of the Roll Angles (front and rear), something we have described before in Lesson Two. This is the Concept of Balance. Now we will work with the roll angles and describe how different spring rates and other settings affect the roll angles to create a perfect setup platform, and see how much of each change affects the setup balance.

The concept of roll angles is useful so we can develop a balanced setup to where the loading on the four tires is ideal and predictable. In some racing circles, this is called the platform, and this would be a good description. In road racing and high-performance formula racing, it is also called the platform and that term includes the aero platform, or attitude that produces the best aero properties.

In this Lesson we will stay with the roll angles and platform as it relates to the setup of the race car. This basic concept is not a new one, but it has been refined and perfected over the past twenty years or so. And, it has been tested and proven to be an accurate and helpful way to achieve a dynamic balance for any race car.

Every race car needs to have a balanced setup. This car has a balance, but also some roll angle to it. There is no advantage to eliminating chassis roll altogether. We can easily configure the body and front valance so that the air is sealed off from coming under the front of the car, even while it is rolling a little. We still need some roll angle and compliance in our race cars. Even when observing Formula One cars, we see that they roll going through the corners. Here we will work with a race car and see how we can change the roll angles.

A Short History Of Chassis Research - In years past starting just after World War II, automotive engineers sought ways to measure the performance and balance of an automobile and later on in the late 1980's and early 1990's, the interest shifted to race cars. A method was developed that analyzed the roll stiffness and what was called the roll couple for a chassis and its suspension systems.

Roll stiffness related to one suspension system and roll couple was the analysis of the stiffness relationship between the two suspension systems in a car. This was a good approach, but for race cars, it was deficient for several reasons we won't go into now. Just know, for any race car to be both fast and efficient and for that performance to stay with the car longer, we need to balance the setup.

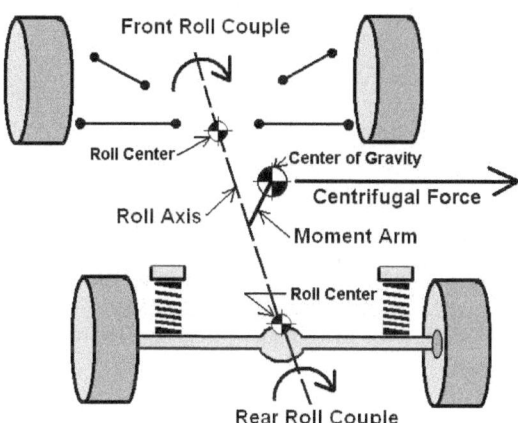

Roll Axis - Roll Couple Distribution Method

The roll axis and roll couple method of evaluating the "roll stiffness" of a chassis was a good start to understanding chassis dynamics. This line of thinking was limited and probably worked alright for a production car, but not so well for many of our current day race cars. In this day and age, we know so much more about chassis dynamics and how to apply methods that will truly tell us if our chassis setup is balanced.

How Is The Roll Angle Produced? – The actual roll angle in a suspension system is produced when the car is cornering and a lateral force is applied to the Center of Gravity of the sprung mass. The sprung mass is the part of the race car that is not a part of the wheels, tires, solid axle, etc. and that is supported by the suspension.

17

This sprung mass wants to keep traveling in the same direction the car was going down the straights before it began to travel into and through the turns. Once the car begins to turn, a force pulls on the car towards the outside of the turn away from the radius point. This is called the Centrifugal force, or what we call CF.

The G-force is the CF divided by the sprung mass weight, or gravitational force on the sprung mass. If the CF were twice the gravitational force on the sprung mass, then the G-force would be 2.0. The magnitude of the G-force helps determine the amount of the roll angle.

As we described in Lesson Three, it is not only the G-force that creates the roll angle, it is the combination of the G-force and gravity which forms the Resultant force and its direction of "pull" on the Center of Gravity.

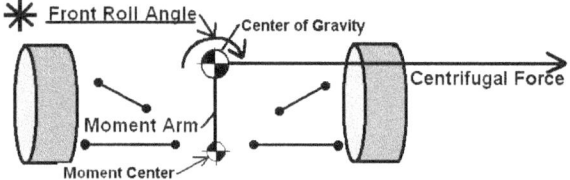

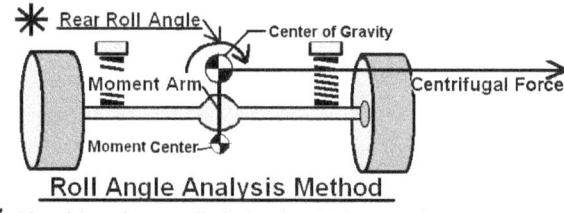

Roll Angle Analysis Method
※ Matching the two Roll Angles Balances the car

The more complete method for determining balance with a setup for a race car is by using what is called the Roll Angle method. This was developed in the mid-1990's and applied to a computer program. The critical dimensions and weights are entered and the program calculates the desired roll angles for the front and rear suspensions. This program was designed for a circle track race car with a AA-arm front suspension and a solid axle rear suspension. We will try some setups to see how we can balance the two roll angles.

What Determines The Degree of The Roll Angle? – So, we have a lateral force which pulls on the CG which produces a roll angle in a suspension system. We can then measure the angle if we know certain things about the suspension system. Here are the things we need to know:

1) The magnitude of the Resultant force (G-force and Gravity combined)

2) The height and width of the Center of Gravity

3) The height of the moment centers

4) The spring base width front and rear

5) The stiffness of the four springs

6) The spring split if one spring is a different rate than the opposing spring

The formulas, or algorithms as they are often called, are very complicated, but we have a tool that can measure the roll angles. We will use that tool to demonstrate how we measure and adjust the roll angles. But what do we use these roll angles for?

How Do We Use Roll Angles? – The process of using roll angles to measure the balance of a setup has been in existence for some twenty years. This process has been used by thousands of race teams, but most race teams still do not know how to use this method, and that is part of the reason you are taking this course, to learn new things.

We calculate the roll angles at each end of the race car and then compare those angles. In most asphalt circle track racing, and road racing with prototypes and formula cars, we match the angles precisely. That is to say, we match them to within a tenth of a degree.

For example, to help you understand what that tenth of a degree means, it is the movement of one wheel by one-eighth of an inch in a system with a 66 inch track width. That is a small amount indeed, but those limits are what we need in order for the system to perform.

In other forms of racing including dirt racing, the approach is very different. Because the relationship of the front and rear roll angles determines the load distribution, we might want to cause the roll angles to be different so that our loads will be biased towards grip for one end or the other.

The end we usually want to have more grip in a dirt car is the rear when the track is very slick. This gives the car a greater acceleration, but at the expense of lateral grip, of which there is very little to begin with. If the gain in acceleration outweighs the loss in lateral grip, then the car goes faster around the track.

For dirt cars, the teams will vary the difference in roll angles depending on the track conditions. Naturally, if the track has a lot of grip, then the front and rear roll angles will be made to be much closer to equal, much like on asphalt with its high grip level. Then as the track changes to become more slick, which often happens, the team will make changes to the setup to cause the rear roll angle to be more than the front roll angle giving the car more rear grip.

How Much Roll Angle Do We Need? – There is generally no perfect roll angle, but there are considerations that differ for different types of racing. Some race cars need to be more compliant than others, meaning that the suspension systems need to move more. So, for those cars, we need for the suspension springs to be softer, like with dirt cars.

For most of today's race cars that race on asphalt, the roll angles are minimal. The goal should not be to eliminate the roll angle, but to limit it for various reasons. Those reasons include the following:

1) To provide a more level and low attitude for stock cars to improve aero properties

2) To reduce camber change in a AA-arm suspension

3) To provide very little vertical movement for a more stable aero platform in both stock cars and prototypes and formula cars

4) To reduce camber change for race cars with wide tires such as prototypes and formula cars

5) In some cases, to increase the heat in the tires for shorter runs

How Do We Make Changes To The Roll Angles? – Because we need to match the roll angles, or miss-match them as the case may be, we need to know how to change the roll angles once we know what they are for our particular race car. We do that by making changes to the following:

The Spring Rate – We can change the overall spring rate, or stiffness in a suspension system. Making the springs stiffer will reduce the roll angle. Making the springs softer will increase the roll angle.

Spring Split - Causing a spring split, or difference in spring rates on each side of a suspension system, will cause a difference in roll angle. For a AA-arm suspension, softening the inside spring rate, or wheel rate for this system, will decrease the roll angle. Softening the outside wheel rate will increase the roll angle. We can stiffen and soften this system by changing the motion ratio, spring angle, or actual spring rate of the installed spring or bump device.

For a solid axle system, the same applies with a difference. That is, the spring base is the distance between the top of the springs, not the wheel rate. For the purpose of roll angle determination, wheel rate has no value. The chassis only knows that it sits on the two springs.

So, for this solid axle system, we can not only change the spring rates and spring split to make roll angle changes, much like the AA-arm suspension systems, but we can also make changes to the spring base width. This is more common in dirt racing where, as we stated above, changes to the roll angles are needed to compensate for changing track conditions.

Moment Center Heights – We can change the roll angles by altering the moment center height. In a AA-arm suspension system, it is harder to do because we would need to make changes to the control arm angles, and that is not a quick and easy change. It can and is done, but not usually at the race track.

For solid axle suspension systems, the changes can be made much easier. If the solid axle is held in place laterally by a panhard, or J-bar, most race cars with this system have slotted mounts on each side so that the bar can be easily raised or lowered by some amount.

If we raise the bar, we shorten the moment arm and the suspension becomes harder to roll. Therefore, we see less roll angle when we raise the bar.

If we lower the bar, we lengthen the moment arm and the suspension becomes easier to roll. Therefore, we see more roll angle when we lower the bar.

How Do We Balance The Setup Using Roll Angles? – Now that we know what influences the roll angles, we need to go to the car and make changes to balance the car. Later on we will tell you how to read your roll angles directly and/or read the tire temperatures to know how to balance the roll angles. We will be making changes to actual race car roll angles to show you how to do it yourself.

For Circle Track Race Cars - In a circle track car using a AA-arm front suspension and a solid axle rear suspension, the front is not easily adjustable and we usually set the front spring rates and bumps based on knowing what type of race track we will be racing on, such as high or low banking. So, the front is not considered an easily adjustable part of the car for at-the-track adjustments in most cases.

As for the rear, the solid axle is highly adjustable in several ways. We can change one or both spring rates, the moment center height, and the spring base if the car was constructed to do that. Let's say we had too much roll in the rear and we need to make changes to reduce it. As we stated above in our explanation on how to make roll angle changes, we could increase both spring rates equally, decrease the LR spring rate, and/or increase the RR spring rate to create a larger spring split. Also, we could raise the panhard or J-bar, or widen the spring base. All of those would decrease the roll angle in the rear.

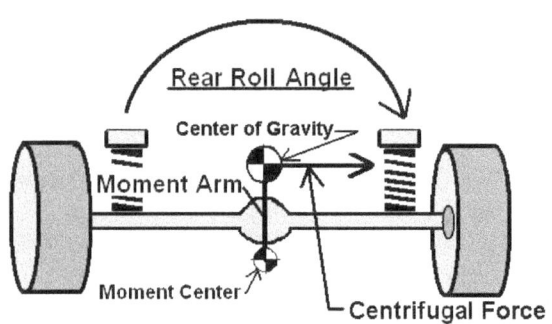

In a typical circle track race car, it is easier to adjust the roll angles to be the same by working with the rear roll angle only and matching it to the front roll angle. The roll angle is much harder to change in a AA-arm suspension. Here we will work with spring rates, spring split and rear moment center heights, i.e. the panhard/J-bar heights.

Which Of The Changes Make The Most Difference? – Now we will tell you how much each change makes for a typical 2800 pound asphalt late model race car with a 15.0 inch high center of gravity, a 1.0 inch high front moment center, and running on a race track that has 12 degrees of banking and where the G-forces are 2.0 G's.

The front spring rates are 1650ppi (pounds per inch) left and right (using a 150ppi ride spring and a 1500ppi bump spring, or bump stop). The LR spring is a 175ppi and the RR spring is a 225ppi for a 50ppi spring split. The panhard bar heights off the ground are 9.0" left and 10.0" right and we are using a 7/8" hollow sway bar. The roll angles are 1.29 degrees front and 2.49 degrees rear. So, in order to match the roll angles, we will need to reduce the rear roll angle. Each of these changes will tell you approximately how much roll angle difference each of the changes makes.

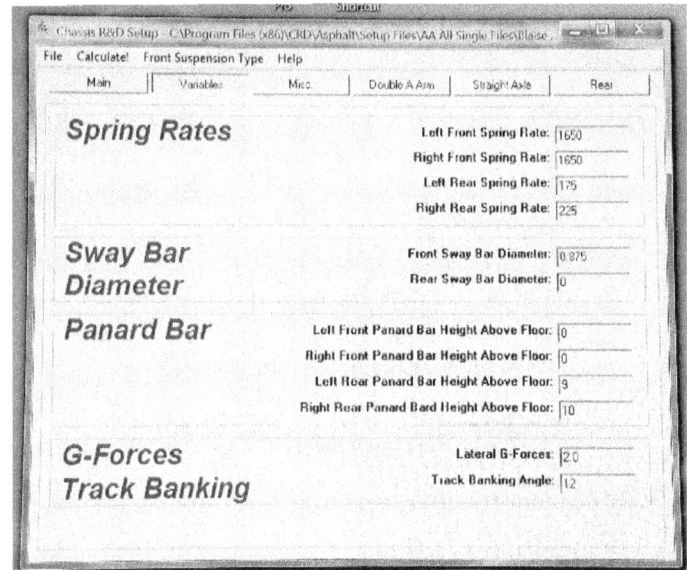

Here is our starting setup. We can see the four spring rates, the sway bar diameter, the panhard bar heights, the Lateral G-forces and the track banking angle. We want to match the front and rear roll angles by making changes to the numbers. Let's see what happens.

When we calculate the roll angles for our base setup, we get a front roll angle of 1.3 rounded off and a rear roll angle of 2.5 rounded off. We need to reduce the rear roll angle in order to match the front roll angle. Let's try some things.

Spring Stiffness – If we first stiffen both rear springs by 25ppi, for a LR spring rate of 200 and a RR spring rate of 250, the rear roll angle changes to 2.32. This is in the right direction, but too little of a change.

If we continue to raise the rear spring rates and not change the spring split, we would need to install a 475 LR spring and a 525 RR spring to get the rear roll angle down to 1.27 degrees. This is not a viable solution because the rear spring rates end up being way too stiff.

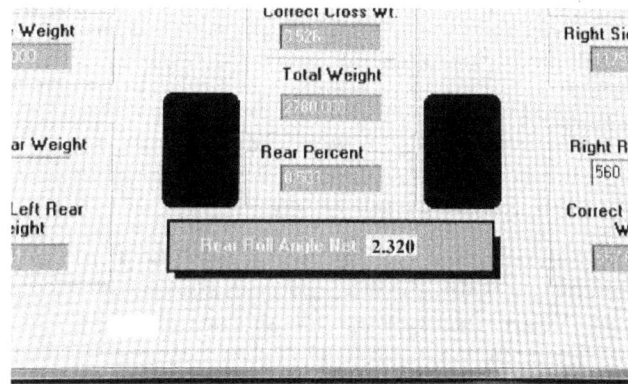

If we stiffen both the rear springs by 25ppi (pounds per inch), we get a rear roll angle of 2.3. This is less than we started out with, but not near enough of a change. Let's keep going.

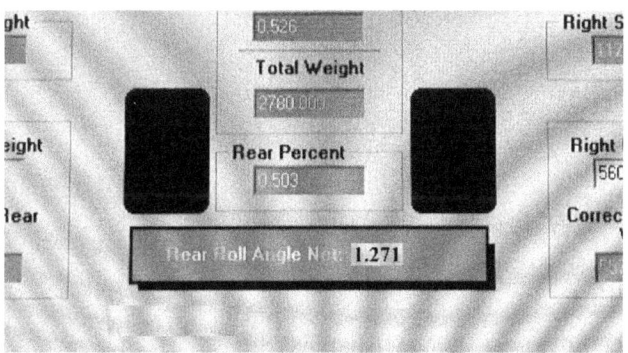

We kept increasing the rear spring rates while keeping the same 50 pound spring split and we had to go all of the way up to a LR rate of 475ppi and a RR rate of 525ppi in order to get to a rear roll angle of 1.27. This is not feasible and we need to find another way to reduce the rear roll.

Spring Split – Going back to our original spring rates, if we now reduce just the LR spring rate by 25ppi, the rear roll angle goes down to 1.92. This was a more significant change, but not enough either.

Going back to our original spring rates, if we then increase the RR spring rate by 25ppi, our roll angle goes to 1.90, about the same as when we reduced the LR spring rate.

If we increase the spring split and then go to 150 LR and 250 RR spring rates, our rear roll angle is now 1.32, or very close to the front and good enough to go racing. But we may not want that much rear spring split.

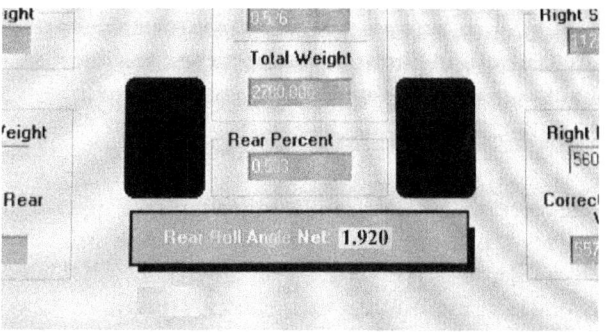

If we go back to the original setup and reduce the LR spring rate by 25ppi to a 150ppi spring, we then have reduced the rear roll angle to 1.9, a more significant change, but not nearly what we need.

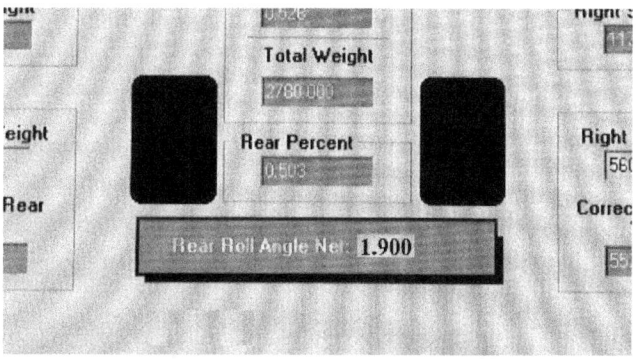

If we go to the RR spring and increase it by the same 25ppi, leaving the other setting the same as the original setup, we get a rear roll angle of 1.90, similar to the change we made to the LR spring rate, but again, not enough.

Panhard Bar – Going again back to our original setup, if we raised the panhard bar by ½" on both sides, we will have raised the rear moment center by that same ½". The rear roll angle is now 2.07, again not enough, but better.

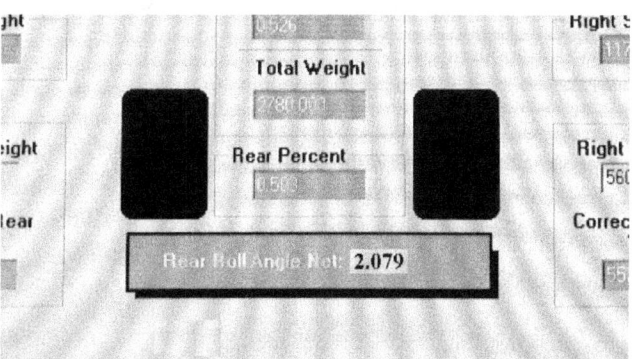

Again, going back to the original setup, if we raise the panhard bar by ½ inch on both sides keeping the same split, we then have a rear roll angle of 2.07, better, but not what we need. Maybe a combination of increased spring split and raising the panhard bar would get us to where we need to go.

Refining The Setup – To get the rear roll angles to match, we can use one or more of these three different methods. We can use spring stiffness, spring split, raising the panhard bar, or some combination of those. Let's see how that works.

If we increase the spring split from 50ppi to 75ppi by installing a LR spring rate of 150ppi and a RR spring rate of 225ppi, and if we also raise the panhard bar by ¾" on each side, our rear roll angle now becomes 1.24 degrees, or just 0.06 degrees different than the front roll angle. We now have a balanced setup.

For a race car with AA-arm front and rear suspensions, the teams work with both spring stiffness and/or sway bar rates to balance the roll angles. In most cases, they want to maintain a certain ride stiffness to help maintain the aero platform and so, changing the sway bar rates offers the best way to balance the roll angles.

should be able to see where we can predict the balance condition of a race car by observing the roll angles.

This method is only as useful as the accuracy of the information entered into the software. There are other ways we can read the balance of the car by reading tire temperatures and we will get into that later on.

Next up we will discuss Weight Transfer and Tire Loading, and how the roll angels can influence those two. We are closing in on a full understanding of how to perfect the race car setup.

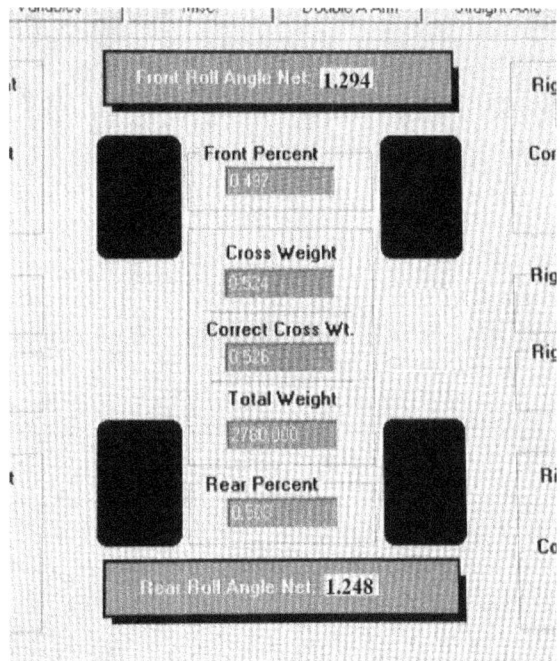

From a starting point being the original setup, we now increase the spring split by reducing the LR spring rate by 25ppi for spring rates of 150 LR and 225 RR. We then raise the panhard bar by 0.75" to 9.75" left and 10.75" right keeping the one inch bar split. Then our roll angles are calculated to be the 1.29 front and 1.248 rear. This is close enough to go racing with. This car now has a balanced setup and it was not that hard to change it to get there.

Summary Of Lesson Four – You have been introduced, or re-introduced if you have come across this method before, to one of the most significant tools we have available to us for setup design. By now you

Exam - In The Context Of This Lesson:

The Platform and Matched Roll Angles Are What?

1) Different in concept
2) The same basic concept
3) Used to perfect shock values
4) Both derived from early development in chassis dynamics

Roll Stiffness And Roll Angles Are Affected By What?

1) Spring rates
2) Moment center heights
3) Spring base
4) All of the above

We Use The Roll Angle Calculations To Do What?

1) Work for zero roll angle
2) Create a balanced setup
3) Improve the ride on bumpy tracks
4) Determine what shocks to use

Which Change Produces The Greatest Difference In Solid Axle Roll Angles?

1) Raising overall spring rates
2) Raising the panhard bar
3) Increasing the spring split
4) Changing the spring base

How Close Do We Need To Make The Roll Angles For Asphalt Racing?

1) Within 1.0 degree
2) Within 0.01 degree
3) Within 0.10 degree
4) It varies with the track conditions

How Close Do We Need To Make The Roll Angles For Dirt Racing?

1) Within 1.0 degree
2) Within 0.01 degree
3) Within 0.10 degree
4) It varies with the track conditions

Race Car Technology – Level Three
Lesson Five – Weight Transfer & Static Loading

Weight Transfer is a process whereby the loading on the four tires of a vehicle, in this case a race car, changes when there is acceleration. Acceleration isn't only a change in forward speed as we commonly know it as, but also when forces are acting on the race cars Center of Gravity in the turns. When we have a change in direction such as going through the turns, that change generates a lateral acceleration as well. We call any acceleration a G-force.

In this Lesson, we will be discussing both lateral and longitudinal acceleration. Both create weight transfer and the re-distribution of loading on the four tires. While braking and accelerating happen mostly on the straighter portions of the race track, the acceleration from cornering is all about a force that is pulling out in a direction away from the radius of the turn. This transfers weight from the inside two tires to the outside two tires.

When we are studying how to make a car handle more efficiently, we are interested in how the tires are loaded. The load distribution on the four tires is what determines how our car is going to handle and how much speed it will be able to carry through the turns. To better understand dynamic loading, we need to first understand how weight transfers and what affects that transfer.

In this example, the car on the inside is keeping all four wheels on the ground, so the inside front tire still has load on it despite the weight transfer. The outside car is lifting the left front tire off the ground, so all of the front weight has been transferred to the right front tire. This is 100% weight transfer at the front. This is basically a three-wheel car and not the ideal situation.

Race cars with a low center of gravity, like these modifieds at Boman Gray Stadium, transfer less weight and therefore retain more inside weight on the tires through the turns. This provides more grip due to more equally loaded tires.

Weight Transfer during cornering takes weight off the left, or inside, tires and puts that weight onto the right, or outside, tires. The amount of weight can easily be calculated, but where the loads end up on the four tires can be very unpredictable. With a balanced setup where both roll angles are matched, the distribution of weight is predictable and can be determined.

25

Formula and prototype race cars are very stiffly sprung and therefore the balance of the setup is more critical as it applies to weight transfer. The slightest miss-match in the roll angles causes a greater change in weight distribution.

What Affects Lateral Weight Transfer? – *Weight transfer happens in all vehicles that are turning. Even vehicles that don't have a suspension will transfer weight. A good example of that is when we corner too fast in a car with a high center of gravity, it tips up onto two wheels. That represents 100% weight transfer, and all of the vehicles sprung weight must be supported by the outside tires because they are the only ones on the ground.*

The following are items that affect the amount of weight transfer:

1) *Weight of the Sprung Mass* of the race car. The greater the weight, the more that is transferred.

2) *Height of the Center of Gravity.* The higher the CG, the more weight that is transferred.

3) *Track Width* – dimension from center to center of the tires. The wider the track, the less weight transfer.

4) *Magnitude of the Lateral Force* measured as speed over change in direction. The more force, or speed and the shorter the radius, the more weight transfer.

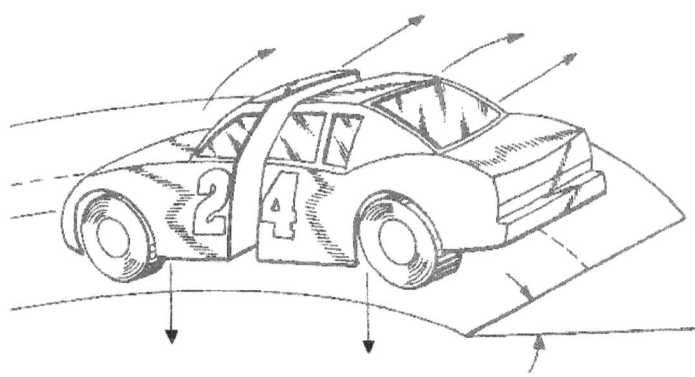

We calculate weight transfer by analyzing the front and rear of the car separately. There is a weight transfer happening at each end. For a car with AA-arm front and rear suspensions, the calculations are fairly simple. For a car with one or more solid axle suspensions, the calculation is different. For the solid axle, there are two calculations, one for the sprung weight and one for the un-sprung weight. These are then combined to yield a total weight transfer at that end of the car.

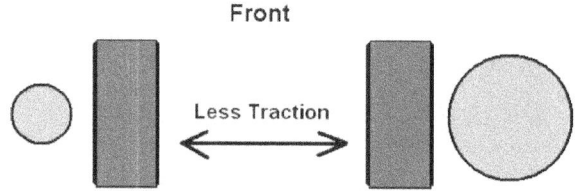

The result of an unbalanced setup where excess load transfers to the Right Front tire causing the rear tires to be more equally loaded. This would be a very tight handling car.

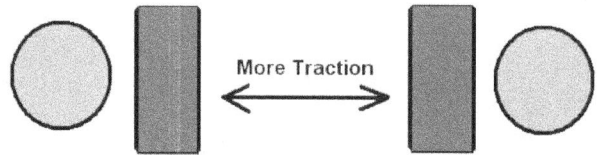

If the roll angles are different with the rear rolling more than the front, then we have excess weight transfer at the front causing the right front tire to do more work and the left front tire to do less work resulting in more un-even loading on the front tires and therefore less grip. The rear tires are more equally loaded and therefore gain grip. Loss of grip in the front combined with the gain in grip at the rear makes this a tight setup, or one that produces under-steer.

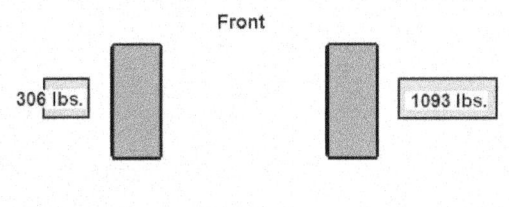

Ideal Dynamic Weight Distribution Has Equal Weight On Both Inside Tires and Equal Weight On Both Outside Tires

The ideal situation is when we balance the setup by matching the roll angles front and rear. Then the weight transfer becomes predictable and if we can arrange the static weights correctly, we can end up with equal weight on the inside tires and equal weight on the outside tires. The level of grip at the front and rear are now the same, if the car has the same size tires on the front and rear of the car.

How Do We Calculate Weight Transfer? - The formula for weight transfer is:

WT = (Sprung Weight x CG x G-force) / Track Width

In this exercise, we are going to calculate weight transfer for a car turning in one direction. We will then back out what we need for static weight distribution, or as we call it in circle track racing, cross weight, in order to end up with the ideal loading on the tires at mid-turn. This is a big deal. You can now know what weight distribution you need to start with in order to have the ideal weight distribution at mid-turn. This will only work if the roll angles are matched up and the setup is balanced.

For this example, we have a race car that weighs a total of 2,800 pounds. We are going to divide that into percentages of front to rear. Since this car is a 50-50% car front to rear, we will have 1,400 pounds resting on the front tires and 1,400 pounds resting on the rear tires. We also have 54% left side weight, so the left side weighs 1,512 pounds (2800 x .54 = 1,512) and the right side the remainder, or 1288 lbs. We'll use those numbers later. Here are all of the numbers we will use:

Vehicle Sprung CG = 15.0"
Front Weight = 1,400
Front Un-sprung Weight = 150
Front Track Width = 66.0
Rear Weight = 1,400

Rear Un-sprung Weight = 300
Rear Track Width = 65.0
Rear Un-sprung CG = 13.5 (this is the height of the axle)

Since we subtract the un-sprung weight from the total front and rear weights to get sprung weights:

Front Sprung Weight = 1,400 – 150 = 1,250
Rear Sprung Weight = 1,400 – 300 = 1,100

We now have to calculate the weight transfer for each end for the sprung weight, and at the rear for the un-sprung weight of the solid axle assembly. The front, or for any AA-arm suspension, un-sprung weight does not transfer. It stays right where it is. Using the formula:

For the Front Sprung Wt /   WT = $\dfrac{1250 \times 15.0 \times 1.5}{66.0}$ = 426.13 lbs. Remember this number.

For the Rear Sprung Wt /   WT = $\dfrac{1100 \times 15.0 \times 1.5}{65.0}$ = 380.76 lbs.

For the Solid Axle /   WT = $\dfrac{300 \times 13.5 \times 1.5}{65.0}$ = 93.46 lbs.

The total rear WT = 380.76 + 93.46 = 474.22 lbs. Remember this number.

Adding the front and rear weight transfers, we get a total weight transfer for the race car of 900.35 pounds. Let's just round that off to 900 pounds. In our example, the left side weighs 1512. If we transfer 900 lbs. at the front, there will be 612 remaining on the left side. The right side weighs 1,288 lbs., so if we add the 900 lbs. of transferred weight to that, we get 2188 lbs. ending up on the right side tires at mid-turn.

Ideal Weights mean that equal weights will end up on the two right side tires and equal weight will end up on the two left side tires. By that rational, we divide the right and left side dynamic weights by two (2) to end up with:

Left Side Dynamic Corner Weights = 306 (612 / 2)
Right Side Dynamic Corner Weights = 1,094 (2188 / 2)

Those weights are what the corners will have on them if we start out with the correct corner weights and weight distribution. Calculating that is easy. If we know the final weights we want, and we do, then we just subtract the amount of weight transfer at each end to know the static corner weights we need to start out with. Let's do it.

RF = 1,094 − 426 = 668
LF = 1,400 − 668 = 732 (this is the remainder once we subtract the RF from the total front weight)
RR = 1,094 − 474 = 620
LR = 1,400 − 620 = 780 (this is the remainder once we subtract the RR from the total rear weight)

Now we have the corner weights we need to start out with at static ride height in order to have the perfect loading on the four tires at mid-turn after the weight has transferred. If we add the RF and LR weights, we get 1,448 lbs. If we divide that by the total weight of 2,800 lbs. we get 51.7% cross weight. If we dial in 51.7% cross weight reading on our scales, we will have the perfect load distribution on our tires starting out to end up with the perfect load distribution at mid-turn, ONLY IF, we have a balanced setup with the roll angles being matched up.

Coincidentally, the program we used to calculate the roll angles in the last Lesson does this calculation and gives the user the correct cross weight that will produce the ideal dynamic loading. But now you can do this yourself. All you have to do is enter your race cars numbers and do the calculations.

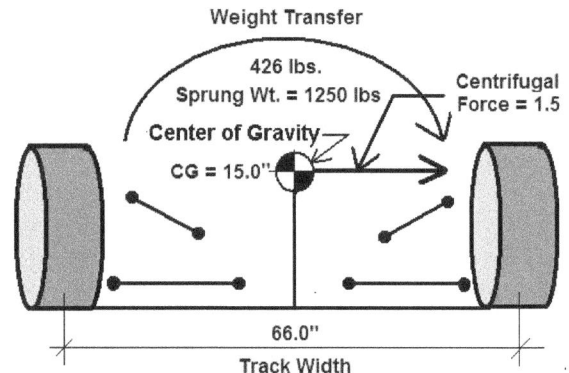

We can calculate the weight transfer at the front, or any, A-A-arm suspension. We need to know the Sprung Weight, the Center of Gravity height, the Track Width, and the lateral G-force. Then we can calculate, using the formula shown, the weight transfer, and then the correct static tire loading that will result in the ideal tire loading at mid-turn.

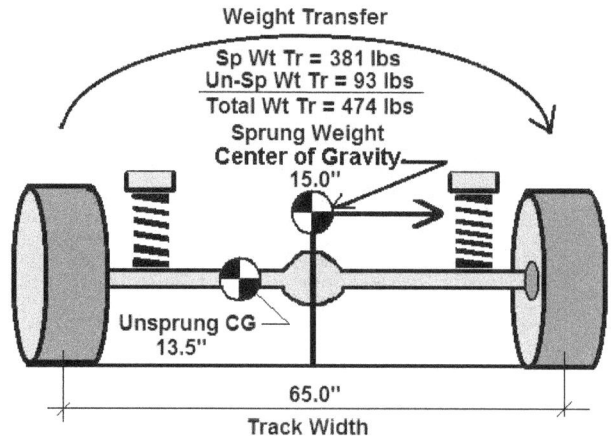

Calculating the weight transfer at the rear of a stock car, or any solid axle suspension, is different than what we do for a A-A-arm suspension. We need to do two weight transfer calculations. One is for the sprung weight transfer and one is for the un-sprung weight transfer of the solid axle assembly. These two are then added together to come up with a total weight transfer at this end of the car.

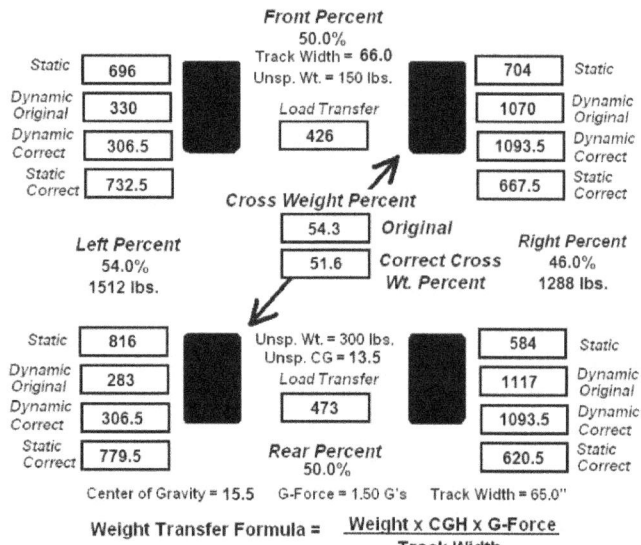

Here are all of the numbers for our Sample weight transfer calculation. We have demonstrated that we can find the correct static weight distribution that will yield the correct dynamic weight distribution after the weight has transferred. For this to work, the setup must be balanced.

How Do Roll Angles Affect Weight Transfer? – The calculations we made to determine the static tire loading that would result in the ideal dynamic loading of the four tires through the turns is only possible if the roll angles are equal. If the roll angles were different, then the weights would end up a different distribution and what that would be depends on how much the angles were different. It would be impossible to calculate what static weights would be needed in order to end up with the ideal dynamic weights if the setup were not balanced.

Plus, an unbalanced race car with un-equal roll angles is very unpredictable. If the banking or G-forces change slightly, the weight distribution changes too because the roll angles will change. The dynamic load distribution would be very unpredictable and so would the handling of the car.

Longitudinal Weight Transfer – Longitudinal weight transfer moves weight from the rear of the car to the front when we are braking. It then moves weight from the front to the rear when we are accelerating. These actions will affect the loading and the grip of each end of the car.

The front will gain a lot of grip from braking and this helps the car turn into the corner. As soon as the brakes are released, the loads equalize again and we are ready for mid-turn lateral weight transfer.

The rear gains grip from accelerating in much the same way it gains front grip from braking. The G-forces are much higher as a result of braking than what is generated from acceleration. How do we know that? It's easy.

The car takes a full straight to accelerate to the speed where the car will begin braking. Then it is a short distance, compared to the length of the straight, to where it again returns to turn speed while braking and decelerating. The short time to go from full speed to turn speed verses the longer amount of time it takes to accelerate to full speed means the G-forces generated have to be much more for braking. Acceleration force is determined by the amount of speed gained or lost over the amount of time. Less time equals more force.

The acceleration that adds loading to the rear tires while the car is accelerating helps those tires gain the grip that is needed to counter the work load the tires are asked to do. They must resist the lateral acceleration caused by the car still turning plus the torque of the motor pushing the car faster. This added grip is used to keep the tires from slipping and losing traction.

Since we are neither braking or accelerating at mid-turn, the Longitudinal weight transfer is of little concern for setting up for the mid-turn balance. But it does affect the corner entry and corner exit.

Just to be complete, there is a portion of the entry segment and the exit segment of the turns where there is a combination of longitudinal and lateral acceleration. On entry, this loads the right front corner a greater amount and on exit, it loads the right rear corner a greater amount.

On braking into the corner, there is longitudinal weight transfer and we can see the front of this car has moved down, probably onto the bump stops or bump springs. Since the longitudinal G-force is greater when braking harder, there is more weight transfer from the rear to the front. This can upset the balance of the car more so than easier braking. Driver instructors will caution against late, hard braking on entry for that very reason.

When the car is accelerating off the corners, weight transfers from the front to the rear. This car shows this effect as we can see how high the front valance is off the ground. When the car was at mid-turn, that same valance was on, or very near, the ground. The weight transfer has taken weight off the front springs and put that weight onto the rear springs. Even if this car had shocks with more rebound to where it would not rise up in the front, the weight transfer would still occur.

So, let's calculate the weight transfer from braking and accelerating. We use the same basic formula for weight transfer that we used for the lateral weight transfer. We just adjust the numbers somewhat.

Here are the numbers we will use for Braking:

Total Vehicle CG = 15.0
Longitudinal G-force = 2.5
Total Vehicle Weight = 2,800
Front Un-sprung Weight = 150
Wheel Base = 110.0
Rear Un-sprung Weight = 300
Race Car Sprung Weight = 2350

A Note About Longitudinal Weight Transfer: In longitudinal weight transfer, the un-sprung weight of the front and rear do not transfer. They stay where they are.

Using this formula:

WT = (Sprung Weight x CG x G-force) / Wheel Base

Because race cars can brake with significant force, especially the formula cars, our G-force number will probably be higher than our lateral force number. We will be using 2.5 G's of braking G-force. Let's Calculate the weight transfer on braking.

$$WT = \underline{2350 \times 15.0 \times 2.5} = 801 \text{ lbs.}$$

Because race cars accelerate at a much lower G-force than when braking, our G-force number we use for acceleration will be lower than the number for braking. We will be using 0.75 G's. Let's Calculate the weight transfer when accelerating.

Here are the numbers we will use for Accelerating:

Total Vehicle CG = 15.0
Longitudinal G-force = 0.75
Total Vehicle Weight = 2,800
Front Un-sprung Weight = 150
Wheel Base = 110.0
Rear Un-sprung Weight = 300
Race Car Sprung Weight = 2350

$$WT = \underline{2350 \times 15.0 \times 0.75} = 240 \text{ lbs.}$$

In these examples, we see how much weight transfers due to longitudinal forces for both braking and accelerating. The braking weight transfer is a lot. The harder we brake, the more G-force we generate and the greater the weight transfer. That is exactly why we cannot brake super hard into the corners. If we do, the weight balance shifts to where we won't be able to control the car.

For acceleration, we can see what we have to work with for distribution on the rear tires. If we transfer 240 pounds when accelerating, then maybe we can move more of that transferred weight to the inside tire to gain more grip.

Summary For Lesson Five – This discussion and the calculations will be most useful for the next Lesson in RCT Level Three. Now that we understand how weight transfers and how we can determine the desired tire loading at mid-turn, we can also calculate the spring force we would need to produce that tire loading. On to Lesson Six.

Exam - In The Context Of This Lesson:

Weight Transfer Is Caused By What?
1) Acceleration of the race car in some direction
2) An un-balanced setup
3) Soft ride springs
4) Gravity acting on the sprung weight mass

Weight Transfer Becomes Predictable When?
1) The spring rates are the same front and rear
2) The setup is balanced
3) The roll angles are matched
4) 2 and 3

What Affects Lateral Weight Transfer?
1) The weight of the sprung mass
2) The lateral G-force
3) The height of the Center of Gravity
4) The track width
5) All of the above

A Setup Is Balanced At Mid-turn When?
1) There is equal weight on all four tires
2) There is more weight on the inside tires
3) More weight on the outside tires
4) Equal weight on the outside pair of tires

Braking Causes Weight Transfer Due To?
1) Lateral acceleration
2) Centrifugal force
3) Longitudinal acceleration
4) A higher center of gravity

Accelerating Off The Corner Does What?
1) Un-loads the front tires
2) Adds load to the rear tires
3) Provides more rear grip
4) All of the above

Lesson Six – Weight Distribution Front To Rear

In our last Lesson, we learned about weight transfer and how to calculate it. We used, for an example, a car with 50/50 front to rear percent of weight distribution. Now let's see what happens when we change the front to rear weight.

The static weight distribution is one of the most important settings on a race car. It can determine how the other parts of the setup will work. Ultimately, we need the perfect weight distribution on the four tires at mid-turn. As the rear percent goes up, so does the cross weight for a circle track race car. We'll show you how that works.

Importance Of Front To Rear Percent – There are many teams who stand by and maintain a certain cross weight percent. Even if that percent worked with the old car and/or old setup, changes to the car might produce a change to the front to rear percent number. And chances are very good that the weight distribution being used is not correct.

If you install a lighter wheel assembly or a lighter rear end unit, you will have changed the front to rear percent and along with that, the relationship of the unsprung weight to the sprung weight. Going from a General Motors engine to a Ford in circle track late model racing can change the front to rear percent because the Ford motor is usually heavier.

If we change the front to rear percent from 50/50 to 48/52, that being 52% rear weight, we will demonstrate how by doing the calculations, the cross weight must be increased so that we will still have equal loading on the side pairs of tires.

If we go to 52% rear weight distribution, then we have 2% less weight on the front tires and 2% more weight on the rear tires. Two percent of a car weighing 2800 pounds is 56 pounds. That doesn't seem like a lot of weight, but when we do the calculations, you'll see that the cross weight percent must go up considerably to compensate for that change.

The static weight distribution helps us achieve the proper tire loads when we negotiate the turns. Every race car needs a given weight distribution that can be calculated easily.

Doing A Sample Calculation With A Higher Rear Percent - In this Lesson, we will be doing the calculations for weight transfer again using a different rear percent than we did in the last Lesson. And, we will show you approximately how the cross weight percent must go up as the rear percent goes up. Conversely, if the rear percent goes down, then the cross weight must also go down.

Since we are using the same total weight, the same unsprung weights, same Center of Gravity height and the same side percentages, etc., our weight transfer numbers will be the same. What will change will be the front to rear weight distribution and the static loading on the tires which will produce a higher static cross weight.

Here are my calculations that are shown on a data sheet I made up.

Weight Transfer Data Sheet

Total Weight __2800__ Front Un-sprung Weight __150__
Lateral G-force __1.50__ Rear Un-sprung Weight __300__
Center of Gravity Height __15.50__ Rear Un-Sprung CG Height __13.50__
Front Track Width __66.0__ Rear Track Width __65.0__

Starting Corner Weights __706__ | __638__
                                   __806__ | __650__
Left Side Weight __1512__ Right Side Weight __1288__
Front Weight __1344__ Rear Weight __1456__
Front Sprung Weight __1194__ Rear Sprung Weight __1156__
Starting Cross Weight __51.6__ Rear Weight Percent __52.0__

Weight Transfer = Sprung Wt x CG x G-force / Track Width

Front Sprung Weight transfer = $\frac{1194 \times 15.0 \times 1.5}{66.0}$ = __407__

Rear Sprung Weight transfer = $\frac{1156 \times 15.0 \times 1.5}{65.0}$ = __400__

Rear Solid Axle Weight transfer = $\frac{300 \times 13.5 \times 1.5}{65.0}$ = __93__

Total Rear Weight Transfer = Rear Sprung Wt Trans. + Solid Axle Wt Trans = __493__

Total Weight Transfer = Front Wt Trans. + Total Rear Wt Trans = __900__

New Left Side Weight __612__ / New Right Side Weight __2188__

Dynamic Corner Weights = NLSW / 2 | NRSW / 2 = __306__ | __1094__
                        NLSW / 2 | NRSW / 2 = __306__ | __1094__

New Corner Wts = Front Wt. - RFNCW | RF Dyn. Wt. – Front Wt. Trans = RFNCW
                 Rear Wt. - RRNCW | RR Dyn. Wt. – Rear Wt. Trans = RRNCW

New Corner Weights = __657__ | __687__
                     __855__ | __601__
New Cross Weight Percent = RF + LR / Total Weight = __55.07__ ←

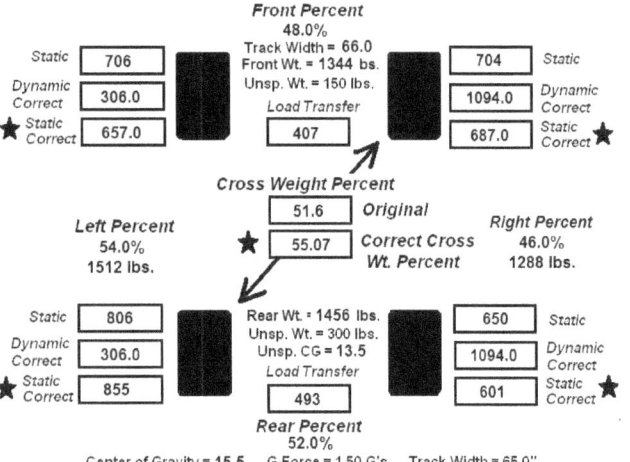

As this data sheet shows, all of our original numbers stayed the same as in Lesson Five, except for the corner weights. This car has 52% rear weight and so the static corner weights are different. What we end up with is a correct cross weight of 55.07 percent. That cross weight, or weight distribution, will give us the ideal tire loads at mid-turn once the weight transfer has taken place.

Some asphalt, and most dirt cars, run a higher rear percent of weight than 50/50. This modified runs anywhere from 50.5 to 51.0 percent. At 50.5 percent, with its numbers, it runs 53.2 percent of cross weight. It won 14 races in a row at five different race tracks in Florida including New Smyrna Speedway 2016.

On flatter tracks, there is more need for bite off the corners and using more left rear loading on the tires helps achieve this. If the car has a higher rear percent, up to a point, then the cross weight will need to be higher and that puts more left rear weight onto the tire. At mid-turn, the loading is normal, but coming off the turns when the car is running more straight ahead and turning less, the left rear load increases the bite for better acceleration.

Road Racing Applications – In road racing, the cars turn both ways. Since we are only turning one way when we are circle track racing, we can choose any static weight distribution we want. For road racing, the cars turn both ways. The cross weight must be the same for both directions or else the car won't handle the same for right and left hand turns. So, these cars must run a 50/50 diagonal weight distribution.

If the cross weight is dependent on the front to rear percent, and we know it is, then as we setup the car for road racing, we then need to change the only thing we can change, and that is the front to rear percent in order to end up with a 50/50 diagonal weight distribution.

For road racing, the diagonal weight distribution must be 50/50. These cars turn both ways and if the distribution were not that way, then the car would handle much differently between right and left hand turns. If the front to rear percent of weight distribution were not correct, then 50% cross might not work for having correct weight distribution through the turns.

For road racing, in order to end up with the ideal dynamic weight distribution, we need to adjust the front to rear percent to a number that needs 50% static cross weight, or 50 percent both ways. That would help the car to handle in a neutral balance no matter which way it is turning.

Let's assume we are taking this same sample circle track race car road racing. First off, we will need to move weight side to side to make the left and right side percent numbers the same, or 50%. We also need to reduce the rear percent. Since the cross weight percent is at 51.6 for our Lesson Five car that had a 50/50% front to rear distribution, in order to get to where it needs the 50% cross, we would have to move weight and go down on our rear percent to arrive at the lower 50/50 weight distribution that would be ideal for road racing.

When we went from 50% rear to 52% rear, we had to increase the cross weight from 51.6 up to 55.07 percent. Going down from 51.6% cross to 50% cross is about half the gain when we needed when we went from 50% to 52% rear percent. So, let's try using 49% rear to see how it goes.

Our calculation is for a left or right hand turn since our side percent numbers are the same. When we go through these same calculations, we end up with very nearly 50% static cross weight in order to have the ideal tire loading through the turns.

Weight Transfer Data Sheet

Total Weight __2800__   Front Un-sprung Weight __150__
Lateral G-force __1.5__   Rear Un-sprung Weight __300__
Center of Gravity Height __15.5__   Rear Un-Sprung CG Height __13.5__
Front Track Width __66.0__   Rear Track Width __65.0__

Starting Corner Weights __714__ | __714__
                                __686__ | __686__
Left Side Weight __1400__   Right Side Weight __1400__
Front Weight __1428__   Rear Weight __1372__
Front Sprung Weight __1278__   Rear Sprung Weight __1072__
Starting Cross Weight __50.0__   Rear Weight Percent __49.0__

Weight Transfer = Sprung Wt x CG x G-force
                  ─────────────────────────
                        Track Width

Front Sprung Weight transfer = __1278__ x __15.5__ x __1.50__ = __452__
                               ──────────────────────────
                                       __66.0__

Rear Sprung Weight transfer = __1072__ x __15.5__ x __1.5__ = __383__
                              ─────────────────────────
                                      __65.0__

Rear Solid Axle Weight transfer = __300__ x __13.5__ x __1.5__ = __93__
                                  ─────────────────────────
                                         __65.0__

Total Rear Weight Transfer = Rear Sprung Wt Trans. + Solid Axle Wt Trans = __476__

Total Weight Transfer = Front Wt Trans. + Total Rear Wt Trans = __926__

New Left Side Weight __474__ / New Right Side Weight __2326__

Dynamic Corner Weights = NLSW / 2 | NRSW / 2 = __237.0__ | __1163.0__
                         NLSW / 2 | NRSW / 2   __237.0__ | __1163.0__

New Corner Wts = Front Wt. - RFNCW | RF Dyn. Wt. – Front Wt. Trans = RFNCW
                 Rear Wt. - RRNCW | RR Dyn. Wt. – Rear Wt. Trans = RRNCW

New Corner Weights = __715__ | __713__
                     __685__ | __687__
New Cross Weight Percent = RF + LR / Total Weight = __49.93__ ←

These are the calculations for a road racing car with the weights changed to equal side percentages and different rear percent. We chose 49% rear weight and used the same total weights, same un-sprung and CG numbers, etc. We ended up with 49.93 percent of cross, or very near the desired 50/50 weight distribution we would need on a road course.

Using Different Cross Weight Numbers – Suppose for example we have a circle track race car that needs 51.6 percent cross weight in order for the dynamic tire loadings to be correct. What if we increase the cross weight percent, then what happens? Let's take the static cross weight up to 56.0%.

When we increase the static cross weight percent, we are increasing the static RF and LR weight. We are also decreasing the LF and the RR weight. Since our weight transfer at each end stays the same, then we end up with more RF dynamic weight and less LF dynamic weight.

At the rear, we end up with less RR dynamic weight and more LR dynamic weight. In this situation, the front becomes less equally loaded and the rear becomes more equally loaded. For our 51.6% cross example, we saw a difference in loading between the left and right side tires of 788 pounds at each end, or the same amount.

In our 56.0% cross weight example, we ended up with a front difference of 912 and at the rear we had a difference of only 664. That means that the rear tires are more equally loaded by 248 pounds and will therefore have more grip. Since more equally loaded tires have more grip, then this car will be tight, or understeer, going through the turns.

What Happens Next? – Typically, if the team does not consider a cross weight change as a fix to the tight setup problem, then they might do other things to make the car more neutral. They might install spring rates that would otherwise make the car loose, but now bring it to a more neutral handling balance.

What if the original springs were the correct springs that would make the roll angles the same and balance the setup? When the team changes the spring rates to make the car more neutral, the setup is no longer balanced. It might be neutral in handling, but not dynamically balanced.

This syndrome is all too common in todays racing. That is why learning about weight transfer and mid-turn tire loading is so important. Setting up a race car to be as fast as it can be and to win does not means just making it neutral in handling. There is so much more to it. And providing the proper load distribution at mid-turn is the way to high performance.

Here Is How It Works – OK, a team puts too much cross weight in the setup. They get to the track and then realize that the car is tight, or won't turn very well. They can do several things to make the car more neutral, other than the obvious which is lowering the cross weight.

First they might raise the panhard bar, by far the quickest and most effective way to make a car less tight. In extreme cases they might raise the RR spring rate, or lower the LR spring rate to increase the spring rate split and loosen the car. Some crew chiefs have been known to roll back the RR wheel by lengthening the RR trailing arm on a three link suspension. This is rear steer and can make the car very loose off the corners.

All of those changes will make the car more neutral, but will take the setup farther away from balanced. The panhard bar/spring changes only help the car in the very middle of the turns On entry and exit, the car will still be tight, but if the driver likes the feel through the middle, he might not complain much, but the car will be slower.

If the grip in the track surface changes, or the driver is forced to run a different line, then the un-balanced setup will show its ugly head and not feel the same. And, the consistency will not be there like it would with a balanced setup. The longer the race is, the more the tires will give up grip and the slower the car will be towards the end.

There is no escaping the problems that incorrect static weight distribution can cause. In the very beginning of this Race Car Technology school, it was explained that performance is about proper loading on the four tires along with having the largest contact patch among other things. The proper loading is right up there at the top of the list.

Summary – Don't be fooled by the brevity of this Lesson. This information is at the very root of making a race car fast and consistent. Along with the roll angle balance, we cannot realize the potential of dynamic balance if the weights are wrong. Now, with the information presented in this and the last Lesson, you can calculate and know exactly what weights to run in your race car.

Next up we will learn how to set our spring loading to the proper force value for static ride height, plus how to read the spring loading so that we can obtain proper tire loading. The possibilities are many and the process is exciting and fun.

Exam - In The Context Of This Lesson:

The Cross Weight That Provides Ideal Mid-turn Distribution Is Dependent On?

1) The total weight of the car

2) The left side percentage of total weight

3) The front to rear percent

4) The center of gravity height

In Our Examples, We Changed The ?

1) The total weight of the car

2) The left side percentage of total weight

3) The front to rear percent

4) The center of gravity height

If We Increase The Front Percent, We Also Must?

1) Reduce the left side percent

2) Reduce the rear percent

3) Reduce the cross weight percent

4) Increase the cross weight percent

When We Increased The Rear Percent, We Also Had To?

1) Reduce the left side percent

2) Reduce the front percent

3) Reduce the cross weight percent

4) Increase the cross weight percent

Road Racing Cars Must Have?

1) Equal side percentages

2) Equal diagonal percentages

3) A front to rear percent that makes 50/50 cross weight

4) All of the above

Race Car Technology – Level Three
Lesson Seven – Dynamic Loading – Force Analysis

Now that we have learned all about weight transfer, we can move into using that information to actually setup the race car. We will now study the art of setup concerning loads and forces. This is where we physically measure the loading on the four tires while the car is on the track.

This presentation represents advanced analysis of what is going on while the car is going through the turns. We are going from predicting and calculating the forces to measuring the actual loads and forces. This is an important step up in our understanding of how to setup a modern day race car.

The loads on the tires at mid-turn can be calculated as well as measured. We will now show you how to use shock and spring measurements to know the spring loads that translate to the wheel loads, and wheel loads to spring loads. This is known as dynamic loading of the four tires. This car uses ride springs as well as bump devices to generate the force needed to support the car at mid-turn.

In some classes, like this asphalt modified, there are no bumps, so the car must use the spring force alone to support the car. For any race car, once we know the wheel force, or tire loading in the turns, we can measure the spring force being generated so that we can use that to produce the ideal wheel load.

In the recent past, the Cup cars were not allowed to use bumps, so they ran the cars into coil bind where the springs compressed all of the way and then the tires became the suspension springs. Tire spring rates of 3,000 to 3,500 ppi were not uncommon. In today's Cup racing, the teams use bump springs surprisingly close to those same rates as the old tire spring rates.

As you travel through this Lesson, it may seem complicated, but in reality, it is more simple and easy to understand than you might think. The calculations we are going through are not necessary for you to use this information on your race car. It is presented to help you understand the importance and need for this approach and to help racers balance the setup.

The methodology presented here has been used with great success for around the past ten years. The process was commercialized about six years ago by Gale Force Suspension. Now, many top teams are using these routines to measure and adjust the setups of their race cars. This lesson will explain not only how it works, but how to properly apply it.

So, let's get started. When a tire carries a load, it must be supported by a spring device in a car with a suspension. Examples of spring devices include coil springs, torsion springs, leaf springs, bump springs, and bump stops. Those are what we see primarily used in race cars in various combinations.

Wheel Rate To Spring Rate – In previous lessons, we have talked about how we can translate the spring rate to a wheel rate. In a similar way, we can translate a wheel force, or tire load, to a spring force. The weight the scales read when we are scaling the car in the race shop must be supported by springs. And those springs must produce a force sufficient to support that scale weight. The key word here is FORCE.

The definition of Force as it applies to this Lesson is this: A spring at rest exerts no force. If we put a weighted object on top of the spring, the weight of that object compresses the spring and then the spring has force. It is holding up that weighted object. If we remove the object quickly from the spring, it would recoil suddenly and return to its original length and again have no force. The force generated is the weight of the object that caused it to compress to a length less than its original length.

So, for a spring under a load, the force is the compressed distance that the spring is verses what it was at rest times the rate of the spring in pounds per inch of displacement. A 200 ppi spring compressed one inch has a force of 200 pounds.

We used the "motion ratio squared" method to translate the spring rate to a wheel rate, and now we can then reverse that process to find the spring force that supports the wheel weight. If, for example, the scale under the right front tire reads 600 pounds, that force of 600 pounds is then divided by the motion ratio squared and that equals the spring force needed to support that 600 pounds.

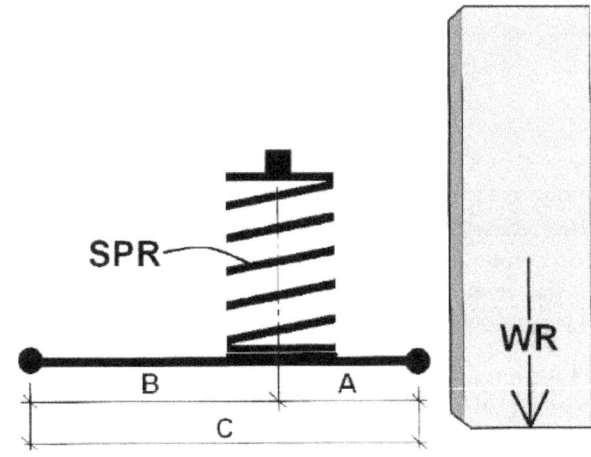

The process to find the motion ratio for a big spring car is the same as we use for a coil-over car. The spring moves at a decimal ratio to the movement of the wheel. To translate spring rates or forces, we use the motion ratio squared.

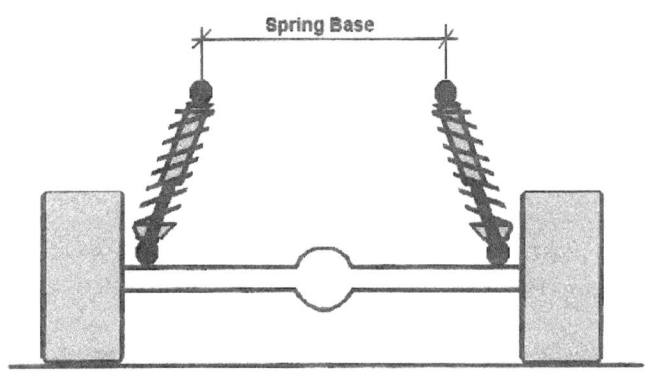

The rear has no wheel rate for the purpose of dynamic evaluation. It does have a wheel force, but that is not easily translated to a spring force. If we know the four dynamic wheel loads that will balance the setup, we only need to make sure one of the wheels, or tires, have the correct loading at mid-turn. That way, the other three loads must be correct too. If one were not correct, then the other three would not be correct either. The total loading of the four tires combined is a finite number and the distribution is dependent on the loads on the four tires.

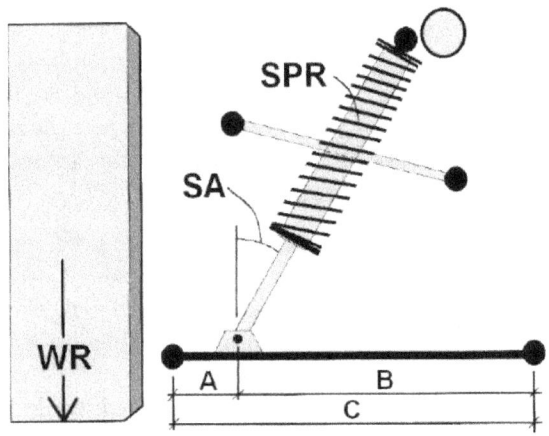

The wheel rate, as we have previously learned about in past Lessons, can be translated to a spring rate. The wheel loads can also be translated in a similar process to a spring force. We use the very same motion ratio. Here is a drawing depicting the measurements we use to determine a coil-over springs motion ratio. We can also just measure the wheel travel verses the spring travel to come up with the motion ratio. When we are translating the forces, we use the motion ratio squared.

For a typical circle track late model, the motion ratio of the RF suspension might be something like 0.8152. Remember that motion ratio is simply the distance the spring moves verses how far the wheel moves. In this case, if the wheel moved 1.0", then the spring moved 0.8152". It's as simple as that and any team member can measure the motion ratio.

If we square that number using a calculator (0.8152 x 0.8152), we get 0.6645. Our scale weight of 600 for the RF tire divided by 0.6645 equals 902.88 pounds of spring force just to hold up the right front corner of the car. And, if that spring were rated at 400 ppi (pounds per inch), then it would be compressed a distance of 902.88 / 400 = 2.257 inches from its free height when out of the car.

If we put that coil over spring and shock on a force measuring machine, and then compressed the shock to the length it was while on the car on the scales, the force should read 902.88 pounds. Compressing the spring to its installed, or measured length, gives us the force it is producing at that length. Remember this process because this is how we can very easily read our dynamic force while on the race track.

For a conventional setup like this one, the shock travel translates into a force that can be translated to a tire load at mid-turn. The only advantage the bump setups have is in lowering the car for a lower CG and better aero properties, and minimal camber change. In a test comparing conventional setups to bump setups, there was only a tenth of a second per lap difference between the two, but a tenth can be the difference between winning and losing.

Reading Shock Travel To Get Spring Force - Suppose we can read the shock/spring travel of the RF corner while the car was on the race track going through the turns. We can do that by using a couple of different methods. We can use a simple shock travel indicator, or we can use data acquisition, both of which measure shock travel. The maximum travel that we see that is due to dynamic forces happens when we are at mid-turn at maximum speed. There could be an influence on this travel from entry braking and that might affect the numbers we see, but we will address how to take care of that later on in this Lesson.

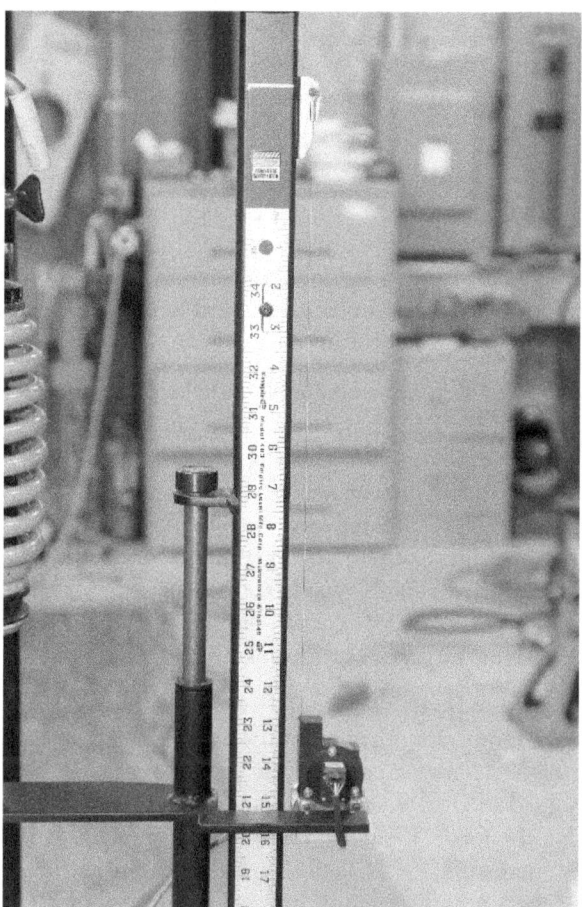

To find the force that the coil-over spring and/or bump produce at mid-turn, we only need to measure the length of the shock while on the car at mid-turn and then compress that shock and spring to the same length in a force measurement machine. On the Gale Force machine, the shock length is shown on the graduated scale in inches on the rig. Once we compress that coil-over to that same length as was recorded on the track, we will then be able to read the force in pounds.

When the car has run some laps on the track at speed, we bring it in and measure the additional shock travel from when it was on pit road at rest and we get 1.859 inches. Since our spring rate was 400 ppi and our spring travel at ride height was 2.257", then if we add those two travels together, we get a total travel of the spring of 4.116". That travel amount multiplied times the spring rate of 400 ppi equals a force of 1,646.4 pounds.

If the spring/shock were 18.5" at free height, then the length at mid-turn would be 18.5 – 4.116 = 14.384". If we put the shock/spring in the force machine and move it to the length, then we could read the force directly. And it would read 1,646.4 pounds, just like we calculated.

That is the spring force we need to hold the RF corner up at mid-turn. Now, we want to know, what is the tire loading? If we multiply that total force number by the motion ratio squared, we get (1, 646.4 x 0.6645) 1,094 pounds. The RF tire supports a load of 1,094 pounds at mid-turn.

To further explain this, if we could put a scale under the tire at mid-turn, that is what the scale would read. Going back to our weight transfer lessons, we can know the RF ideal loading by doing calculations. In Lesson Five, our calculations yielded an ideal RF tire loading of 1,094 pounds for the example car.

Since we match our intended RF tire loading as evidenced by the spring force we are reading, then we have proven that this car is setup to be balanced and that means that the roll angles must be matched. We have now demonstrated how we can measure the balance of the setup by knowing the spring force. We simply read the shock travels and translate that force to a wheel load at mid-turn and then compare that load to what we know is correct for this car.

The tire loads are interdependent. That means, if one tire load not ideal, then all of the tire loads will be wrong. That also means, we only have to measure one tire load to know the entire picture, or layout, of the other tire loads. If the RF carries 100 pounds more load than necessary for a balanced setup, then the LR tire will also carry 100 pounds more, and the LF and RR tires will each carry 100 pounds less than necessary.

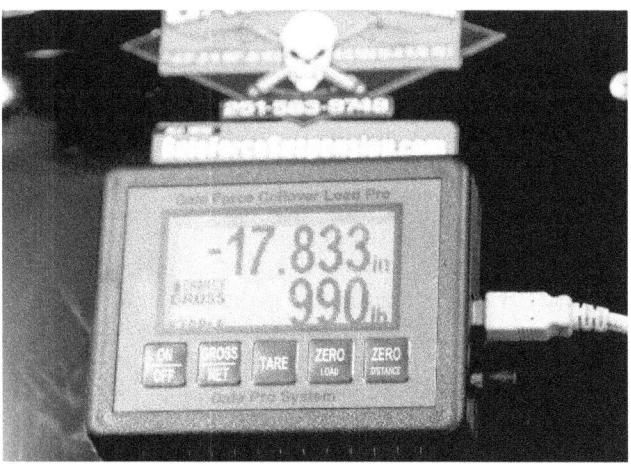

This readout from the Gale Force machine shows us that at a length of 17.833 inches, the force of the spring, and maybe the bump too, measure 990 pounds. If this was our example car, then the wheel load would be 657.86 pounds, or our motion ration squared times the spring load (0.6645 × 990 = 657.86 lbs.).

Modern Day Soft Spring Setups With Bumps – The example car we looked at above was what we call a conventional setup. It uses only ride springs to support the car. Many modern day race cars use what we call bumps in addition to the ride spring. Bumps can be a coil spring with a consistent and linear spring rate, or some form of bump stop, or compressible material with a variable spring rate.

When running on bumps, we can also see when the shock body contacts the bumps and then read how the force increases as the travel further compresses the bump stop. For bump stops, the rate of increase in force, or load, is not linear. It rate of increase rises as the bump compresses. On a graph it produces a curved line. With a bump spring, the line showing the increase in load is a straight line.

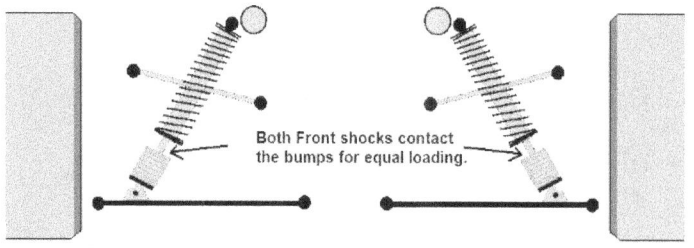

The bumps shown here in red support additional loads that if supported by the softer ride springs would cause the chassis to travel more than it has room to travel. Without the bumps, the car would be hitting the race track. We do the calculations to show you how that works.

This shows a typical bump stop. The initial force needed to compress this particular bump is not a lot. As the bumps is compressed, it resists the compression more and more and therefore the spring rate increases. At some point, it becomes "coil-bound" so to speak and cannot compress any more.

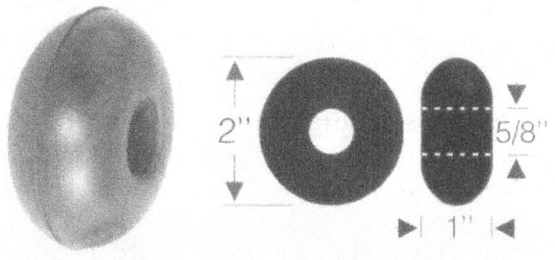

This much stiffer bump device is used on smoother tracks to reduce overall shock travel. These can be stacked in any numbers to change the progression of force and regulate the travel.

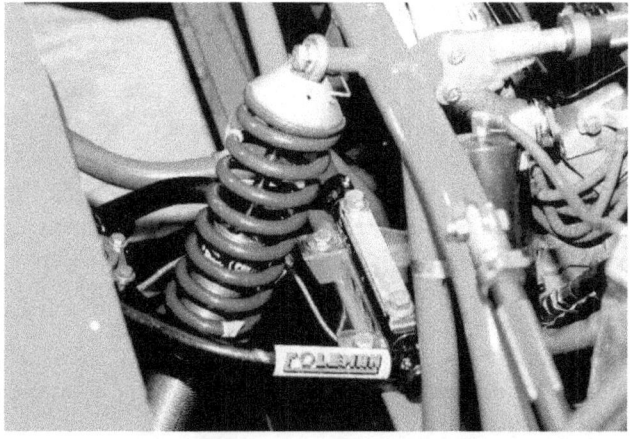

When soft springs are used as ride springs, they must be pre-loaded in order to provide the necessary force to support the car at static ride height. Once on the track, we then use bumps to stop the movement of the chassis to prevent the chassis from bottoming out on the race track.

The bump devices actual spring rate, when activated, are added to the ride spring rate. The bumps are placed on the shock shaft and when the shock moves in compression, the shock body contacts the bump device and the spring rate of the suspension goes from that of the ride spring to a rate equal to the bump spring rate plus the ride spring rate.

If for example, we are using a ride spring with a rate of 150 ppi and a bump spring rate of 1500 ppi, then the spring rate of that suspension when the shock has traveled onto the bump is 1650 ppi. Once the shock comes off the bump, the rate then goes back to 150 ppi.

A note about the bump stops. The nature of bump stops is this. The spring rate of a bump stop increases as it is compressed, and if compressed enough, will go up to a very high rate or even solid. The same is true of the bump spring when it goes into coil bind, or at the maximum travel. But the bump spring rate stays consistent all of the way up until it goes into coil bind. We call this linear progression.

Just as with the conventional setup, we can record the shock travel and then duplicate that in a force machine to read the spring force at mid-turn. In a bump setup, the force is a combination of the ride spring force and the bump force.

The bump spring is used to control the amount of shock travel as does the bump stop. It will compress and provide the force needed to support the chassis at mid-turn. Its force is added to the ride spring force that comes from both the shock travel and the pre-load forces.

Reading Spring Force Conventional Verses Bump – We might think there is a difference in the force readings between the conventional setups and the bumps setups just because the bump combined spring rate is so high. That is not true. The tire loads that are produced by a balanced setup are the same whether the car is running ride springs only, or ride springs plus bump springs.

The only advantages that the bumps setups have over the conventional, ride spring only, setups is that the bump setup produces a lower attitude that creates a lower center of gravity, possibly more aero downforce, and minimal camber change due to limited vertical motion. All of that was covered in earlier Lessons in Race Car Technology.

As to aero downforce, if the front of the car produced an additional 200 pounds of downforce due to the lower attitude of bump setups, then that 200 pounds would be equally distributed between the two front tires and our RF tire load would go up by 100 pounds. This gets into the finer points of reading tire loads to help us to produce a balanced setup, but for now, let's stick to the basics.

Bump Setups Using Soft Ride Springs – If we do the calculations for spring travel to produce the force necessary to carry the mid-turn loads, then with the softer ride springs used on bump setups the travel would exceed the available suspension travel. So, the teams running those setups must pre-load the softer ride spring. Pre-load means to compress the spring while it is on the shock with the shock fully extended. This puts force on the spring without any movement of the shock. Its rate remains the same, but a lot of the force it will need is already there.

If we pre-load that 150 ppi ride spring on the shock by four inches, it would have a pre-load force of 4 x 150 = 600 pounds. If we subtract that 600 pounds of pre-load from the 902.88 pounds of force we need to hold up the car at normal ride height, we end up with a shock/spring travel of 2.0192" at normal ride height. This is what teams actually do when working with relatively soft front springs.

So, we have a mid-turn spring force, from previous calculations, of 1646.4 and a force to hold the car up at ride height of 902.88. The additional force needed to support the car at mid-turn is 743.52, or the difference between the static force and the dynamic force.

If our ride spring is 150 ppi, then 743.52 / 150 = 4.96". Taking that travel out to wheel travel means that the wheel would move 6.08 inches. There isn't that much travel left in the suspension if we set the ride height at 4.0". The car would hit the track big time. We need another spring to work with that force so that we don't make contact with the track. We need a bump spring or bump stop. In this example, we'll use a bump spring.

So, the wheel would move 6.08" with just the ride spring, and we have maybe 3.5" of travel available to stay off the track by 1/2". Doing the math, 6.08 – 3.5 = 2.58". 2.58 x 150 = 387 pounds of force we need to add to the suspension to hold it up off the track.

If we install a 1,500 ppi bump spring onto our shock, then it would travel 387 / 1,500 = 0.258 inches, or about a quarter inch, to create the force needed to keep the car the 1/2 inch off the race track. But if it travels 1/4" under that load, we must subtract 1/4" from 1/2". Now our car is 1/4" off the race track. And that is what we would normally do. We would install a bump device that would create a much stiffer spring rate to hold the car off the race track.

That bumps spring, or bump stop, doesn't have to be a 1,500 ppi spring rate. It just has to develop a force necessary to hold the car off the track. If it were a 3,000 ppi spring, then the travel would be half the 0.258 inches, or 0.129, or about an eighth of an inch.

The Influence Of Braking On Shock Travel – We would be leaving something very important out of this discussion if we didn't cover what could influence the RF shock travel, or the shock travel at any corner of the car. For dynamic load analysis, we need to make sure that the shock travel readings we are seeing are truly created from the dynamic forces the car experiences at mid-turn.

For example, if we have very little anti-dive in the RF suspension, and/or pro-dive in the LF suspension, then on braking into the corner at most short tracks, the loads the RF and LF suspensions will show might be wrong.

The loading in the RF and LF under braking is usually more than we would see at mid-turn. So, the reading we take from shock travel might be wrong. If the shock travel was 0.25" more from braking than what it is at mid-turn, then with a combined spring rate of 1650 ppi at the RF, we would be recording a load that is 412.5 pounds more than is actually what the mid-turn dynamics produces. In other words, the shock travel method is useless if we allow the shock/spring to move that far under braking.

There is a solution. Back in the day, teams used Anti-dive to help reduce the nose dive while under braking to help reduce the camber change that motion produced. In recent years, teams have reduced the use of Anti-dive and even introduced Pro-dive into the LF suspension. Both of those changes have allowed erroneous readings to occur.

It is now time to reintroduce Anti-dive on both front suspension systems. We won't be reducing dive because we'll already be on the bumps, but we won't be overloading the bumps from heavy braking and then the readings will truly represent the mid-turn shock/spring travel. Then we will be able to rely on our shock/spring travel amounts knowing they represent only mid-turn forces.

You have just discovered the secret of using bumps to setup a race car. In conjunction with the soft ride springs and stiff bumps, we use shocks that will work with the bump spring rates and not the ride spring rates. They will be much stiffer on rebound than the ride springs and that keeps the front suspension riding on the bumps much longer. We'll get into the design of shocks for various types of setup in a later Lesson.

Sway Bar Loading And Force Influence – The sway bar is a spring and adds a force to the ride and bump spring forces to assist in holding the car up at mid-turn. At ride height, with no pre-load, the sway bar produces no force, but it still has a spring rate. When the car dives and rolls, the sway bar may be forced to twist, and at that point, it produces a force that adds to the ride spring and bump spring forces.

In the previous examples of forces in a conventional and bump type of setup, we purposely left out the influence of the sway bar so that the explanations wouldn't be overly complicated. And the addition of the sway bar force is not really complicated either. It just needs to be included in the overall consideration for force needed to hold the car up at ride height.

As you would expect, the sway bar force is directly related to the spring rate, or stiffness, of the sway bar system, and the amount the chassis is twisting the bar, just like what happens with spring travel. The more travel, or twist, the more force that is produced. Many designs of late model circle track cars will cause the sway bar to twist when the chassis dives without any roll at all.

Gale Force Suspension makes a sway bar force measuring tool that will measure the sway bar force, if you know the suspension travel at mid-turn. With this tool, you can re-create that movement and then read the sway bar force at the shock/spring position. This provides a force number we can include in the overall force that is needed to support the car at mid-turn. Once we know the sway bar force, we then subtract that force from the force we need to hold the car up at mid-turn. The remainder force is what the ride and bump spring must produce.

Here we see the Gale Force sway bar force measurement tool. When we replicate the suspension travel to what it is on the race track at mid-turn, we can isolate the sway bar force that acts like a spring. This force is then included with the ride spring and bump forces to make up the required RF corner force needed to support and balance the car at mid-turn. The tool is mounted where the shock/spring is mounted, so the sway bar force is measured at the same place as the coil-over spring is measured at. Then it can be directly added to the spring/bump force.

Since the sway bar can easily produce 150 to 300 pounds or more of force, it is a serious consideration when we are trying to develop the total suspension force to support the car at mid-turn. If we don't include the sway bar force into our suspension force calculations, then we would end up with much more spring/bump force than we need.

Using The Sway Bar As A Spring – In modern asphalt circle track late model racing, the use of bump devices includes running a bump on both sides in the front suspension, or only running a single bump, usually on the left side. When running the bumps on both sides, the sway bar acts mostly in roll.

For setups where the bump is only installed on the left side, the size of the sway bar must be increased dramatically in order to provide a sufficient Right Front spring rate to hold up the car going through mid-turn. Whereas with a two bump system that can more efficiently run a much smaller sway bar, with the single bump system, we need much more spring rate from the sway bar and need to install a much larger sway bar. Sizes ranging from 1.50 inch diameter up to 2.00 inch diameter sway bars is not uncommon. We will further discuss the use of a sway bar as a spring in Lesson 14.

Summary – We hope that this has been a big learning experience for those of you who have not been aware of how bumps actually assist us in setting up a race car. We have calculated the ideal loads we want to see on our tires. We translate those loads to a spring load we need to produce the wheel loads at mid-turn. And then we decide to run bumps with soft ride springs, so we then need to preload the ride spring to produce nearly the force needed to keep the car off the track at mid-turn. We then add in bump stops, or bump springs to provide the added force to finish the job.

This whole tire loading/spring force theme represents just one of the three primary elements that serve to create the overall grip in our race car. Next, we will learn all about element number two, the Angle Of Attack.

Exam - In The Context Of This Lesson:

A Spring Has A Force When?

1) Preloaded on a shock

2) It is a loaded sway bar

3) It is compressed

4) All of the above

A Tire Has A Force When?

1) The chassis is at ride height

) The race car is at mid-turn

3) The suspension is on the bumps

4) All of the above

The Ideal Load On The RF Tire Is?

1) When that load equals the RR tire loading

2) When that load equals the LF tire loading

3) When that load equals the LR tire loading

4) When that load equals the spring and bump forces

The RF Tire Load Can Be Translated To?

1) The RR tire load

2) The RF suspension spring load

3) The LF suspension spring load

4) The LR tire load

To Calculate The Spring Force From The Tire Load, We?

1) Use the motion ratio

2) Use the motion ratio squared

3) Use the spring/shock travel

To Calculate The Tire Load From The Spring Force, We?

1) Use the motion ratio squared

2) Use the spring/shock travel

3) Add the spring force, bump force and sway bar force

4) All of the above

Bumps Are Primarily Used To?

1) Provide more force

2) Increase the spring rate of the suspension

3) Keep the suspension from bottoming out

4) All of the above

The Following Can Interfere With The Force Readings

1) Excess camber

2) Heavy braking

3) Acceleration off the corners

4) Running softer springs

Race Car Technology – Level Three
Lesson Eight – Angle Of Attack & How Tires Generate Grip

The following is a basic repeat of the second Lesson in RCT Level One. Some of you might have skipped Levels One and Two, so we needed to make sure you understand this very important concept. It was presented early on in the RCT courses because this concept first of all has never been fully explained before and you are now seeing it again because it fits right in and is a major part of understanding the three main ingredients of how tires develop grip.

One of the first things we talked about in Level Three was loading on the tires. We told you how weight transfers and how it is distributed on the tires going through the turns. We explained how equally loaded tires offer the most grip for a pair of tires on the same "axle". We went on to tell you how a balanced setup will offer the best weight distribution possible and why.

But, here is the clincher. Loading of the tires is not what makes them develop grip to resist lateral forces when we are going through the turns. A car going straight can be easily pushed to the side with relatively little force and it goes there because it has no resistance to that force even though it may have a lot of loading.

An example is the PIT maneuver used by law enforcement to stop a car that is involved in a chase. The police officer goes to the rear quarter panel and moves over against the offending car and pushes it until it spins out. The rear of the car seems to go sideways fairly easily. Its tires have no AOA initially, and so they cannot resist the lateral force that is the police car.

What the tire needs to resist lateral forces is Angle of Attack. Some might call it slip angle or some other terminology, but we will call it by a name that better explains the concept, and that is angle of attack, or AOA for short.

The concept of AOA as we explain it has never been fully examined as far as we know. It is a simple, but effective, description of why a car can drive through a turn at high speeds. It is one of those items of new knowledge that we promised the veteran racers they would get when considering enrolling in the RCT Courses.

One of the key ways, and actually a necessity in, a tire developing grip is from AOA. A tire running parallel to the path of the arc of the curve cannot generate the Grip it needs to maintain its direction along the arc. It would "slip" off the arc, hence the name "slip angle". It is the angle the tire needs to keep from slipping off the curve.

How The Steering Creates AOA - Every race car in the world is a front-end steering car. That is, the driver turns the car using a steering mechanism that is used to change the direction and angle of the front wheels. The rear wheels seem to follow along. But why do the rear wheels stay on course? The are not being steered.

The driver creates an angle of attack for the front tires so they can develop the side force necessary to change direction and also to stay on the track. AOA is needed to counter the lateral force we know as the centrifugal force that is created once the change in direction has taken place. Without this Angle of Attack, the car would not turn no matter how much the tire is loaded, or no matter what the size of the tire contact patch is.

Rear Tire AOA Study - OK, so we have learned how the front end turns and develops AOA, but what about the rear tires? We cannot steer them, setting aside the effects of mechanical rear steer that will be covered later on. How do the rear tires develop the angle of attack needed to stay on the arc of the turn? The answer is: they already have their angle of attack, and here's why.

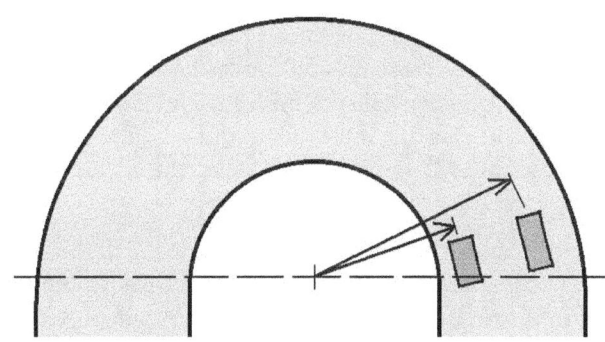

In this illustration, we show how the inside and outside tires have different radii and must follow different arcs when driving through the turn. A line perpendicular to the line extending to the radius point is called a Tangent. To develop the Grip needed to counteract the Centrifugal force, all of the tires must be pointed to the left of the Tangent line.

In the last figure, you will see how each tire follows its own arc. The arc radius of the inside (inside of the turn) tires is less than that of the outside (outside of the turn) tires. In the next figure, each tire is placed over an

47

exaggeration of the arc lines with a radius equal to the distance from each set of tires to the common radius point of the arcs.

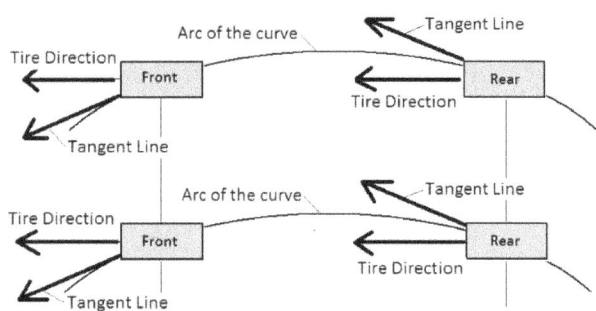

This sketch shows how when the four tires are pointed straight ahead, they form angles to the tangent of the curve that are much different. The front wheels need to be turned left to reach the direction of the tangent line, and more to turn the car. The rear wheels are already pointed left of the tangent line and therefore already have some degree of AOA.

The front and rear tires are pointed in different directions in relation to the Tangent of the arc they are following. In the front, the tires are pointed away from the Tangent and in the wrong direction needed to turn the car.

We must steer those tires left in this left turn in order to not only track parallel to the Tangent line, but a bit more to develop an Angle of Attack needed to generate the amount of Grip necessary to keep the car on the course.

At the rear, we have a different situation. With the rear tires positioned straight ahead, they are already pointed left of the Tangent line and have some Angle of Attack.

If the AOA angle in the rear is less than what is needed, the rear of the car will slip out until the tire angle is enough to provide the sufficient AOA needed to stay on the course. If the angle is more than is needed, then the driver will need to turn the steering wheel more to compensate and to make the car turn.

The front tires react the opposite if they slip. The angle of attack decreases and the steering must be increased in order for the car to stay on the course. We will tell you more about steering angles and front AOA later on. There are limits to how much angle the front tires can be steered to.

When you come to understand this concept of how the car is able to make it around the turns, it will answer a lot of questions people might ask about vehicle dynamics. I have often wondered, given that we have a lot of setup tools to make the front ends turn, why and how does the rear turn with the front? Now we know.

Here we can actually see the lines of the arc of the turn. The front wheels are steered to create the Angle of Attack needed to develop the force to counter the G-lateral forces. The rear tires are already pointed to the inside of the turn and have their own Angle of Attack. This is why both the front and rear tires develop the grip needed to turn the corner.

Angle Of Attack, How Much Is Too Much? There is a limit to how much AOA a tire will tolerate. We know from tire testing that a tire will gain lateral loading when the AOA is increased, up to a point. At some predetermined angle, and it is different for different designs of tires, the lateral force the tires provide goes away. They will stop generating lateral force and then slip.

The design goal for creating maximum lateral acceleration, or grip, is to find the angles that are just less than the angle where the tire loses lateral grip. If we can make the front and rear tires work at that maximum AOA, then the car will be the fastest it can be through the turns.

And, know this, every race car, no matter the design or classification, is subject to the concept explained above. Farther along in this Course, we will get into various design effects, goals and geometry for the rear suspension where we will explain how you can change and fine tune the angle of attack for the rear tires. We can introduce rear steer to gain AOA in the rear in measured amounts.

Now we have covered two of the three elements of how tires develop grip. We have covered loading and now Angle of Attack. Next, we'll discuss tire contact patch optimization. It is the third and final element of grip. Like the two elements we have already covered, without the contact patch, all of the loading and all of the AOA in the world will not maximize the cars mid-turn performance. We must have all three of these working to maximum capacity.

Exam - In The Context Of This Lesson:

Which Element of Grip Is Essential Above All Others?

1) Tire Loading
2) Tire Compound
3) Angle of Attack
4) Contact Patch Area

How Do We Make The Front Tires Develop Angle of Attack?

1) Add More Camber
2) Add Tire Pressure
3) Drive Slower
4) Turn the Steering Wheel

How Do We Make The Rear Tires Develop Angle of Attack?

1) Add More Camber
2) Add Tire Pressure
3) Drive Slower
4) We don't need to, they already have Angle of Attack

How Do We Adjust The Lateral Grip For The Front Wheels?

1) Lengthen the Tie Rods
2) Change The Speed
3) Change The Tire Cambers
4) Turn the Steering Wheel Either Way

How Do We Adjust The Angle of Attack For The Rear Wheels?

1) Change The Tire Stagger
2) Change The Tire Pressure
3) Adjust The Speed
4) Through rear steer geometry

Lesson Nine – Advanced AA-arm Geometry & Contact Patch Optimization

The third and final of the elements that make up overall grip in a race car is contact patch optimization. If we have designed the setup so that the forces on the tires are optimum, and the angle of attack of the tires is sufficient to create a resistance to the lateral forces going through the turns, then we now need a large contact patch.

It is a fact that a tire with a fixed loading and angle of attack will generate more side force to resist the lateral forces during cornering when the contact patch is as large as possible. And, the optimization of the contact patch size comes from tire camber and tire pressure.

Our goal is to find the correct camber, and camber change that will go along with the correct tire pressures to create a larger contact patch during cornering. And all of that is dependent on the tire we are using.

In circle track racing for instance, it is generally known and accepted that Hoosier tires are constructed with a softer sidewall than a Goodyear tire. A team cannot run the same cambers for one tire as the other because of how the sidewall reacts to the camber. Other brands of tires have their own characteristics for sidewall stiffness.

The camber design for race tires has evolved in order to maximize the contact patch area. We will show you how to create ideal camber change and contact patch size for a modern day race car.

Camber Change – As the race car goes into and through the turns, the chassis will dive (or raise up) and roll. These two motions combined are what can cause camber change. If the chassis did not change its height along the centerline, then the roll of the chassis would change the camber by roughly the same angle as the roll angle.

As the chassis dives, and most race car chassis run lower through the turns, there is a "dive" camber change for the outside wheel that is opposite of the roll camber change. In many race cars, we can design the control arm angles so that there is very little, or no, camber change for the outside tire through dive and roll.

The inside tire and wheel will lose camber in most race cars, and with road racing cars, the static camber is opposite of what is needed for the direction we are turning. So, it is beneficial to run as little camber as possible in road racing cars so that the contact patch is somewhat useful for the inside tire.

Tires Don't Like Camber Change - In most cases, race tires do not like camber change. When entering a turn, the tire takes a set, or rolls over and begins to develop Grip. If the cambers do not remain stable, the tire will take more time to adjust, and frankly, we don't have a lot of time for that.

The result is loss of Grip and loss of speed as the tires struggles to find Grip. So, in modern race car design, we need to plan out our AA-arm geometry so that there will be minimum camber change. How do we do that? There are several ways to accomplish the goal of minimum camber change for both the inside and outside tires.

First, we can arrange the control arms with angles so that with the dive and roll we expect to result from going into and through the turns, will result in minimum camber change. This has been done with great success in what we call conventional setups with both stock cars and some formula cars.

Another way to achieve minimum camber change is to reduce the amount of dive and roll. The setups in Formula One cars are very stiff and the chassis does not move vertically or roll very much. This reduces the amount of camber change. The reverse angle of the upper control arms in a F1 car (meaning that the inside chassis mount is higher than the ball joint) may be beneficial and cause a small amount of beneficial camber change.

The third way we can minimize camber change is through the use of what we call bump stops. In modern race car design, some chassis are designed to travel a predetermined amount and then the shocks contact a bump material and basically stops in both dive and roll motions.

These bumps can be made of hard rubber, polypropylene or actual steel springs. In each case, when the shock body contacts the bump device, the vertical chassis motion is reduced to very small movements and the tire cambers change very little while the suspension is on the bumps. The reduction in camber change is one of the primary reasons why the bump setups are beneficial and help the car to turn.

Bump Setup Camber Change – With the advent of the bump setups in circle track and other forms of racing, camber change of any significant amount while on the bumps is a thing of the past. Even so, there is still a lot of camber change going on from when the car is on the grid to when it is going through the turns. That camber change can cause other setup changes we might not easily see.

The tires provide better grip when the cambers are not changing and will stay very close to ideal due to very little vertical chassis movement. But what about the transition onto the bumps? What happens during that process? Few teams consider what happens then. The associated movement can cause other problems.

As we have discussed above for a more conventional car, when we change the cambers, we are also changing the corner heights and along with that, the distribution of loading on the four tires. If the chassis travels some three or three and a half inches down onto the bumps, where does the wedge, or cross weight, go to then? It has to change from where it was statically based on the fact that the tires change their height relative to the chassis.

Weight change at one corner of the car due to the jacking effect of camber change (and that is essentially what it is, the tire jacks up, or down depending on the change) also changes the loading on the four tires.

We already know that the extreme travel associated with bump setups, from ride height to on-track height will also most times pre-load the sway bar. This adds cross weight to the setup. This weight distribution change must be taken into account with the initial setup when we set the cross weight and sway bar preload.

Some Examples Of Camber Change – It is time to run some experiments setting the car up with various cambers and control arm angles. In our first example, we will setup the front end geometry with low upper control arm angles, meaning they are close to level.

This car is a road racing car with the arm angles being equal on both sides for the upper and lower control arms. This car will lose negative camber at the outside tire (car is turning left in this example) and gain negative camber with the inside tire.

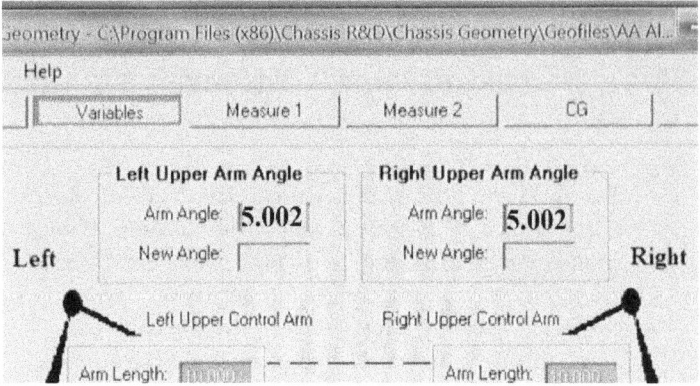

This road racing car has equal upper control arm angles and they are only 5.0 degrees off of horizontal. We'll see how the cambers change when the car turns left and dives and rolls.

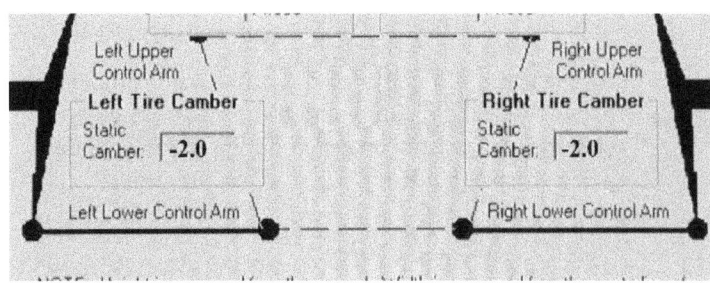

The cambers for the front wheels are equal too because this car must turn both ways. We have a negative (-) 2.0 degrees for the static cambers on both sides.

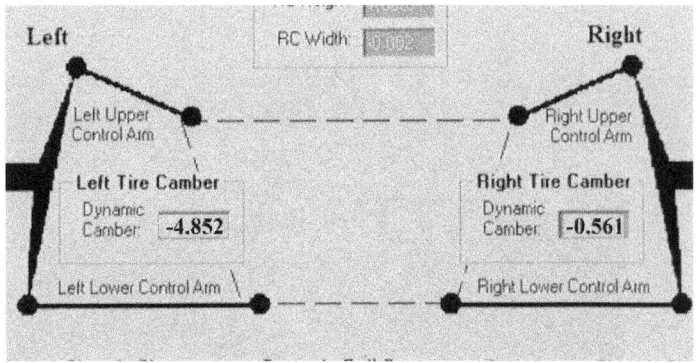

When we calculate the dynamic cambers by entering 1.0 inches of dive and 3.0 degrees of roll, we end up with a left tire camber of -4.85 degrees and a right tire camber of -0.56 degrees. We lost about 1.5 degrees of camber at the right front, or outside tire, and gained 2.8 degrees of negative camber at the inside, or left front tire.

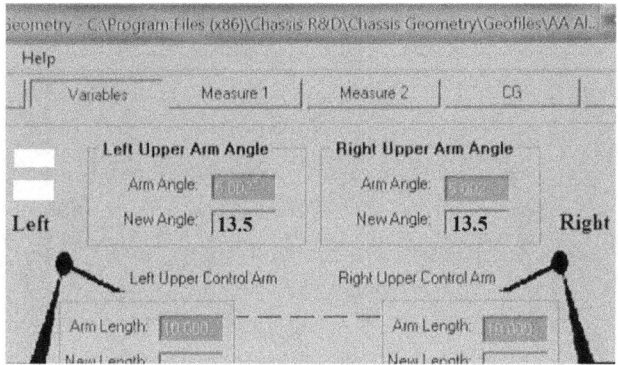

Here we added angle to the upper control arms and now have 13.5 degrees for both upper control arm angles. Look at the "New Angle" box. This should improve, or lessen the outside tire camber change.

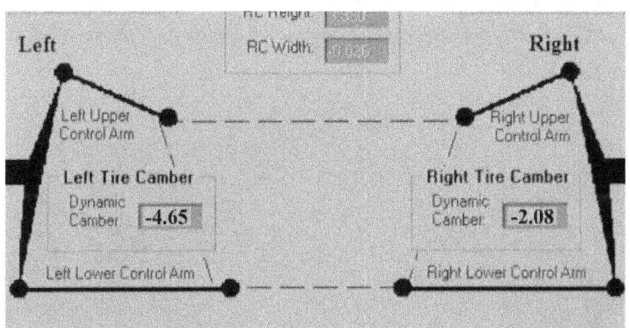

The new dynamic cambers show a -2.08 degrees for the outside tire, or basically what we started with, and a -4.65 degrees of inside tire camber. Our outside tire camber did not change at all and the inside tire camber stayed basically the same as before. The zero change in the outside tire on this car will improve how it turns.

Now let's look at a typical circle track race car running a conventional setup. This will simulate a Tour Modified because this is what we see a lot with those cars. It could still be a late model car, and the results speak to the importance of controlling the camber change.

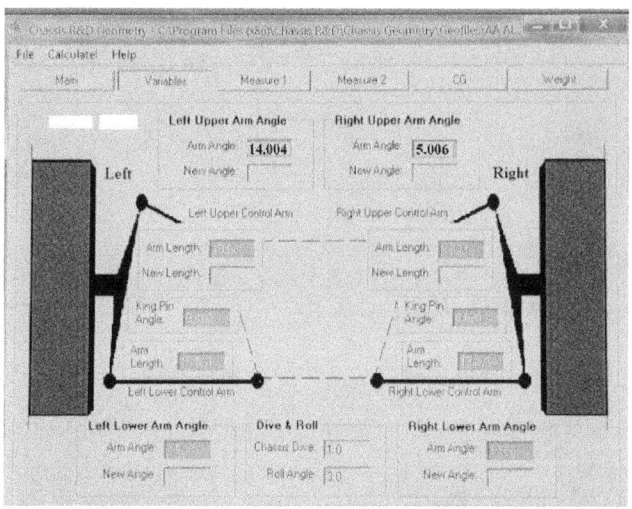

For this example, we use a circle track car that could be a Tour Modified or a late model running a conventional setup. The upper control arm angles are 14.0 degrees for the left upper and 5.0 degrees for the right upper. We also used a common set of dive and roll numbers that relate to more conventional setups of 1.0 inches of dive and 3.0 degrees of roll.

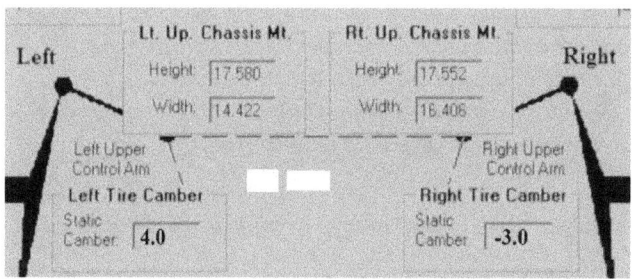

The static cambers for this car are a positive 4.0 degrees on the left front and negative (-) 3.0 degrees on the right front. We'll calculate this and see how our cambers come out.

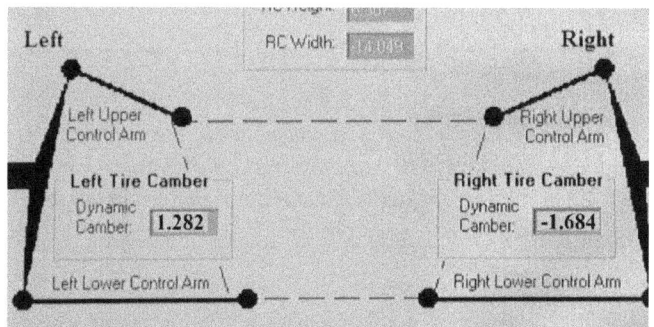

When we calculate the cambers with those upper control arm angles, we get these numbers. The RF lost about 1.4 degrees of negative camber. The LF ended up with 1.3 degrees of positive camber, a number that is not bad at all. But the loss at the RF will be detrimental to how the car turns. Let's see if we can fix that with new upper control arm angles.

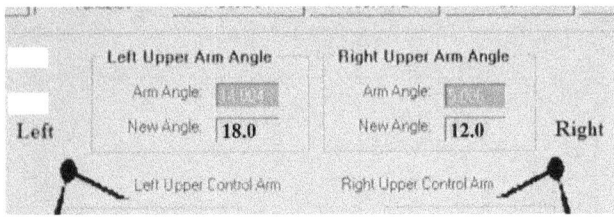

We enter new upper control arm angles and they are 18.0 for the left upper and 12.0 for the right upper. These should improve the camber change, especially at the RF.

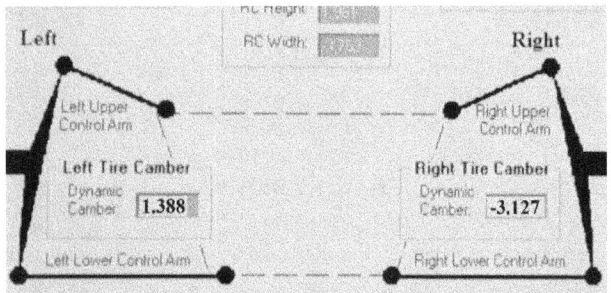

The new cambers show a negative (-) 3.1 degrees at the right front, nearly the same as the static camber, and 1.39 at the left front, nearly the same as before. The fact that the right front camber, or outside in this example, never changed through dive and roll means that this tire will take a set and then provide grip all through the entry and middle of the turn. This provides a much better contact patch to give the car more front grip.

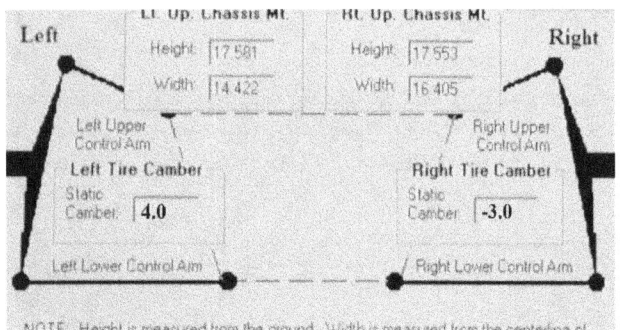

Now we need to look at a typical bump setup. We'll start out with the same front cambers in this example. We will change the dive and roll numbers to simulate the travel onto the bumps and the low roll angle the bumps provide.

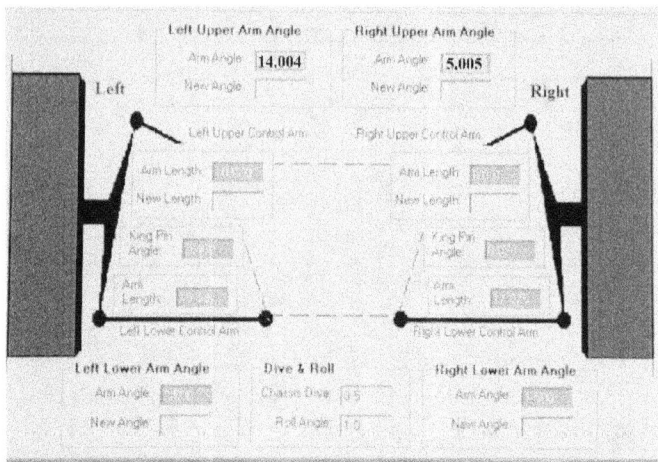

Because the front travels so much with the Bump setups, most teams will put low upper control arm angles into the front end. Here we have 14.0 degrees at the Left upper arm and 5.0 degrees at the Right upper arm. Our dive is 3.5 inches and the roll angle is 1.0 degrees.

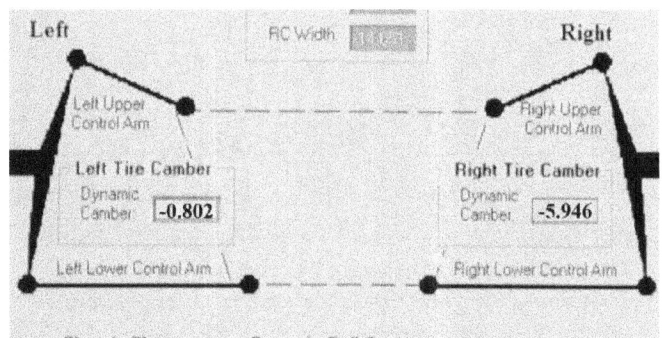

When we calculate the camber change using the upper control arm angles and the dive and roll that simulates the bump setups, we get a RF camber of negative (-) 5.9 degrees and a LF camber of -0.80 degrees. We obviously need to take static negative camber out of the RF wheel and put more positive camber into the LF wheel. But maybe there is a better way to set this up using much different upper control arm angles? Let's take a look.

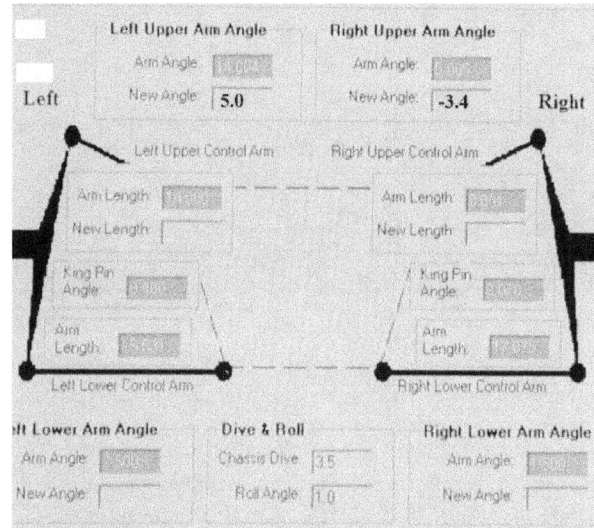

Let's experiment with upper control arm angles and go unconventional. We put 5.0 degrees in the left upper arm and a negative (-) 3.4 degrees in the right upper arm. This means that the right arm chassis mount is higher than the ball joint, which is unconventional. But this car will travel down onto the bumps right away once it is on the track and will really never leave that attitude. We also use the same dive and roll numbers, 3.5 inches of dive and 1.0 degrees of roll.

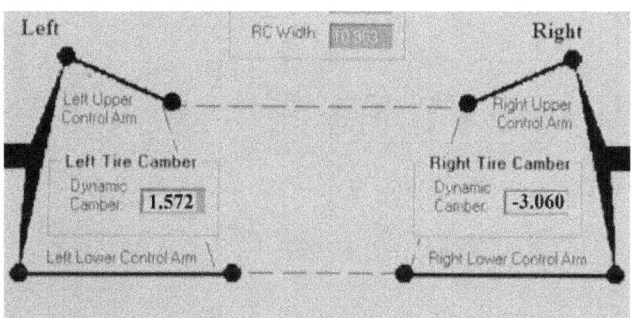

Once we calculate the cambers, we end up with a positive 1.5 degrees at the LF wheel and a negative (-) 3.0 degrees at the RF wheel. These numbers are very close to what we need. So, instead of using a very high positive LF camber and a very low RF camber, we just need to set the car up with more normal static cambers to end up where we need the front cambers to be for the best tire contact patch.

These examples are intended to show how camber change works in the real world. We're not sure anyone has installed exactly the same upper control arm angles as the example, but with what we know about what the tires like, and what happens with the weight distribution when the cambers do change, we might consider the benefits of the examples.

We introduced the concept of weight distribution changes when the camber change, so let's take a look at how that happens. It concerns the height of the spindle related to what camber the tire and wheel are set at.

Weight Change From Camber Change - When we change the static front cambers, or when the cambers change with the car on the race track, we also might be changing the height of the chassis at that corner and we need to bring the weight distribution back to where it was before the camber change. The illustrations show how this happens.

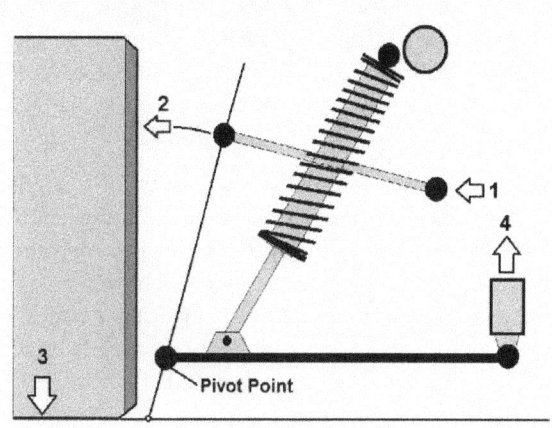

On the left side, we can see here how the intersection of a line through the centers of the ball joints, upper and lower, intersects with the ground inside the tire contact patch. Not all suspensions are setup exactly this way, this is just a demonstration of approximately how this works. If we add positive camber by moving 1, 2 moves in the same direction. Then as 2 rotates around the Pivot Point, 3 moves down lifting 4 the chassis up. This adds more of the total weight of the race car to this tire. We would need to adjust the spring length to bring the chassis back to normal ride height. On the track, we cannot do this, so there is load change happening when the cambers change on the race track.

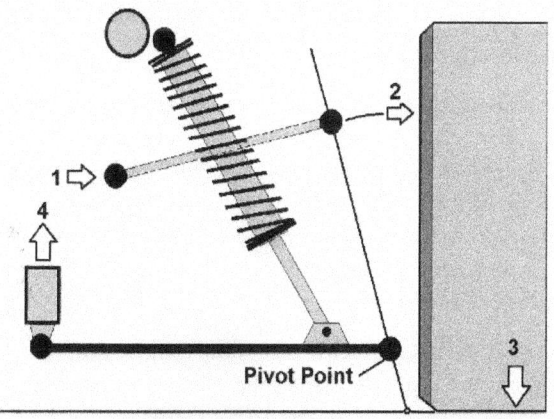

On the right side, what we are seeing is that as we reduce negative camber, or move the wheel towards positive camber, we are also raising the chassis and adding cross weight percent if we don't adjust the spring length to compensate for the chassis height change. In a suspension where the camber changes through dive and roll, no only does the tire not have a consistent contact patch, it also starts carrying a different load.

The concept of weight distribution caused by camber change means that we need to strive for minimal camber change for not only the purpose of contact patch optimization, but also to reduce the change in weight distribution on the four tires.

In the past when I thought about all of the cars I designed and re-designed, the best and most successful cars were the ones where the camber change for both tires was minimal. In todays racing for both conventional setups and bump setups, the same is true as evidenced by the above examples.

Modern Day Contact Patch Shapes - In the above discussion, we learned that certain control arm angles in a double A-arm suspension will create minimal camber change and help us end up with the ideal tire contact patch for a particular race car. We also have discovered through testing and evaluation that a larger contact patch provides more traction in a tire, much like increasing the loading, but not needing to.

Early on we talked about what the tire needs for camber. We are talking about a AA-arm suspension and for stock cars, that is the front tires obviously because there is little we can do to adjust the rear tire cambers for a straight axle suspension. Each brand of tire has different sidewall stiffness and tire construction for the tread support and so need different cambers for each type of tire so that we can end up with the largest contact patch possible.

Like we also pointed out, you may be able to extract a larger contact patch from one tire brand than any other tire brand that has the same tread width and sidewall height. Tire temperatures can point us in the right direction, but we are finding that the driver may be the best litmus test.

As for tire temperatures, a softer sidewall tire can be cambered so that the tire surface temperatures on the side closest to the inside of the turn are 20-25 degrees hotter than the surface temperatures on the outside away from the turn. This difference is achieved with higher camber settings which tend to cause the inside of the tire to flatten out.

Lower tire pressures help us accomplish the goal of producing a larger tire contact patch, but we are not advocating low tire pressures that compromise safety. We do know the trend in racing is to run as low a tire pressure as we can get away with and this is an indicator of the importance of increasing the contact patch area.

In more conventional times, and when using ultra-wide tires, this might be the contact patch that results from having equal tire temperatures across the face of the tread. If we attempted to put too much camber into this wide tire, then we would lose contact loading and area at the outside (meaning outside of the turns) edge of the tire. For more conventional sized tires, this does not represent the ideal camber we need for maximum contact patch area.

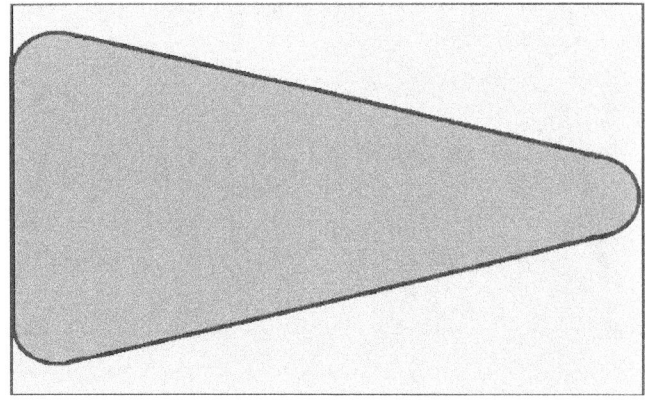

If we add camber to the outside tire and reduce the pressure in that tire, our contact patch might well look like this. The inside (inside of the turns) part of the contact patch is much wider while the outside part is very close to what we showed in the previous sketch. The area of this contact patch is much larger than if we set a camber angle that resulted in even tire temperatures across the face of the tire. For this example, we usually see 20-25 degrees hotter at the inside edge of the tire than we see at the outside edge.

The above example of contact patch resembling a triangle could be applied to this formula car, only in reverse. The outside tires are going into positive camber in this high speed turn. It is possible that the triangle contact patch shape is reversed from what we need for circle track cars. The end result is the same, a much larger contact patch. In this example, this team can set the inside tire camber much more straight up increasing its contact patch size too.

Summation – When you purchase a race car, it is yours. What you do with it is entirely up to you and your team. If you think you need to make changes to the control arm angles based on what we have learned here, then by all means do it. I have personally cut up many hundreds of race cars so that the geometry, mostly for camber change, is better and to help the tires gain contact patch area and with that, grip.

The front of a stock car is hard to tune for dynamic balance, but easy to tune for contact patch optimization. Then when we get the front end as good as it can be, we can easily tune the rear suspension to match the front for grip.

A race car where the front is deficient has to be de-tuned at the rear so that the car will handle neutral. This car will be slower due to lack of overall grip compared to a car that has been optimized.

Exam - In The Context Of This Lesson:

We Need Camber Because?
1) The Tire Distorts When Lateral Force Is Applied
2) The Suspension Points Move In Dive And Roll
3) Camber Helps Generate More Lateral Force
4) All Of The Above

Camber Change Occurs Because?
1) The Wheel Moves Up and Down
2) The Car Rolls
3) The Tire Distorts From Lateral Force
4) A Combination of 1 and 2.

What Changes To The Suspension Will Affect The Amount of Camber Change?
1) Changing the Spring Rates
2) Installing different length control arms
3) Changing the angles of the control arms
4) All of the above

How Much Camber Changes Do We Need?
1) Enough to get what the tire needs
2) An amount equal to the tire roll over
3) As much as possible
4) As little as possible

What Are Common Ways To Minimize Camber Change?
1) By running very stiff setups
2) Through special design of control arm angles
3) With the use of bump devices that limit suspension movement
4) All of the above

The Ultimate Cambers We Need Are Determined By?
1) The Amount of Lateral Load
2) The Degree of Camber Change In Our Suspension
3) The Stiffness of the Tire Sidewall
4) All Of The Above

Camber Change Affects The Tire Loading?
1) True
2) False

Even Tire Temperatures Across The Tire Is Always Optimal?
1) True
2) False

Race Car Technology – Level Three
Lesson Ten – Jacking Force Concepts and Uses

Jacking Force (JF) is a concept that was introduced about ten years ago in an attempt to more accurately define basic vehicle dynamics for a AA-arm suspension. It was thought by some to be the primary force that generates and regulates the chassis dynamics in those suspensions. It is not. It is, in fact, in most cases with modern race car geometry, a system that produces an Anti-roll effect, much like a sway bar.

In this Lesson, we will explain how JF works, how it is calculated, how we might use it to our advantage, and how it might be miss-used. It is extremely important to understand this subject mainly because there is so much dis-information being taught that is floating around the internet. Further, the concept of JF is sometimes made more complicated than it really is. And some of the information is just plain wrong.

Modern day race cars running on circle tracks try to utilize Jacking Forces to create an Anti-roll effect very similar to what a sway bar does. This helps keep the front of the car more level to the racing surface and helps improve the aero effect.

History Of Jacking Force – The effect we call Jacking Force has always been around, but, in production automobiles, the control arm angles are small, so the industry paid little attention to it historically. As we will see in this Lesson, when the control arm angles are small, there is very little JF present in the system.

In the past as far as race cars are concerned, the control arm angles were more than production automobiles, but not sufficient to cause much JF effect. More recently, we see designers and race teams using much greater upper control arm angles with the desire to create higher Jacking Forces.

There are various design reasons why the higher JF is desirable, but know that we only really became concerned with JF over the past decade or so. I myself have a history of discounting JF as a significant effect mainly because the cars I was working with had very little JF incorporated in the design of the geometry. Then I did some experimenting and began working with a unique scale model that could test the JF concept for AA-arm suspensions. Now I can tell you with confidence how JF is produced and how to calculate it.

How This Lesson Came About - In creating this Lesson, we first conducted a series of tests using a unique scale model of a double A-arm suspension whereby we could apply forces to the contact patches and the Un-sprung mass, change the control arm angles, and then measure the roll stiffness and ultimately the chassis roll angle.

The roll angle of the chassis demonstrates the roll stiffness related to sum of all of the effects present in the system and differences in roll angle between the different arrangements of control arm angles tells us how much influence JF has on the dynamics of the suspension.

As a direct result of this testing, we were able to develop roll angle calculations combining the overturning moment method and the JF methods. These calculations resulted in almost exact replication of the actual roll angles we recorded on the scale model.

What Produces Chassis Roll? – The JF does, or can, have an effect on the amount of chassis roll, but what causes chassis roll in the first place? It is what we call the Overturning Moment that is the primary cause of chassis roll. That is, when we enter and negotiate a turn, a lateral G-force is created which tries to pull the car to the outside of the turn. The tires resist this force.

This lateral force causes some weight to transfer from the inside tires onto the outside tires. The amount of weight that transfers is dependent on how fast we go through the turn, i.e. the amount of G-force, the length of the Moment Arm and how wide the track width is at that end of the car, track width being the center to center distance between the two contact patches of the tires. The formula for calculating weight transfer is:

$$WtTr = \frac{MA \times Sprung\ Wt \times G\text{-}force}{Track\ Width}$$

Real numbers might look like this for a modern super late model asphalt car:

$$WtTr = \frac{15.0 \times 1280 \times 1.50}{66.0} = 436 \text{ lbs.}$$

If the chassis is sprung, the chassis will roll as this weight transfer takes place. As the weight transfers, the left front corner of the car looses 436 lbs. and the right front corner gains 436 lbs. The suspension springs for each side must adjust to the new weights, and so the left side rebounds and the right side compresses. If our wheel rates were say 250 lb./in. (340+/- ride spring rates), then each side would move 436 / 250 = 1.74 inches. If we take the ArcTan of [(1.74 x 2) / 66.0], we get an actual roll angle of 3.02 degrees for this car. For a bump setup using 2,000 lb./in. bump springs, the wheel rate would be about 1,520 lb./in. and the roll angle would then be much less, or about 0.50 degrees.

The above is what the chassis would roll to if we don't include the sway bars Anti-roll properties, or the JF properties. Each of those, the sway bar and JF, produce Anti-roll effects that reduce the chassis roll. If we are trying to balance the chassis' front and rear suspension systems, then we need to know something about how much reduction in roll angle each of these Anti-roll effects has. If not, we may think we have a balanced setup when in fact we do not.

A chassis running 52% cross weight and 56% left side percent has a LF static weight of 756 pounds and a dynamic LF weight after weight transfer of 320 pounds. The RF has a static weight of 644 lb. and a dynamic RF weight after weight transfer of 1,080 lbs.

These weights are used to calculate the lateral forces at the contact patch and at the chassis Center of Gravity. We simply multiply the corner weights times the lateral G-force. If the lateral force changes, so does the amount of weight transfer (wheel loads) and so do the Jacking Force numbers. Keep this in mind as you travel from track to track. When your lateral G-forces change, so do the JF amounts. And, If the chassis roll angle changes, so to do the dynamic control arm angles and therefore the JF numbers.

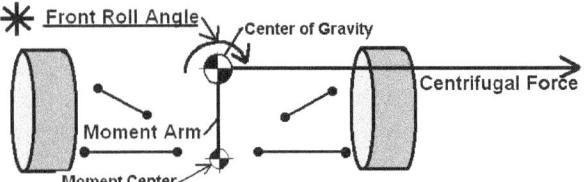

In this illustration, we see where the Centrifugal force acts on the Center of Gravity of the sprung mass and creates chassis roll. This is referred to as the Overturning Moment that causes chassis roll. In a race car, this roll is somewhat restricted by sway bars and possibly Jacking Forces.

Two Types Of Jacking Force – There are two distinctly different types of JF. These two types combine into a single force that can enhance or restrict chassis roll. The first is what we will call Contact Patch Initiated JF because the initiating force comes from the contact patch resisting the lateral forces created when cornering. The second type we will call the Un-Sprung Mass Initiated JF because it is initiated from the weight, or mass, of the un-sprung components in the suspension.

Contact Patch Initiated JF - The first type is a product of the forces acting on the chassis and the tire contact patches of the outside (right in a left turn) and the inside (left side in a left turn) tires that produce a lateral force on the control arms. If the control arms are mounted at some angle from horizontal, then there will be as a result a vertical force acting on the chassis. These forces will add to, or subtract from the chassis roll.

In Illustration 1, we see that as the lateral force is pulling to the right, the contact patch is resisting that force in the opposite direction. This tries to rotate the spindle around the lower ball joint. This motion puts a force into the upper ball joint in the opposite direction of the contact patch force. This force is proportional to the distance from the lower ball joint (the point of rotation) to the ground verses the distance from the lower ball joint to the upper ball joint.

Then, if the upper control arm is mounted at an angle from horizontal, then vertical forces are created. If the angle is down to the centerline, or the chassis mount is lower than the ball joint, then the vertical force would be down onto the chassis.

If the chassis were rolling, then this force is counter to the rolling action and the roll angle is reduced. If the angle of the control arm were reversed with the chassis mount higher than the ball joint, then the roll angle would be increased.

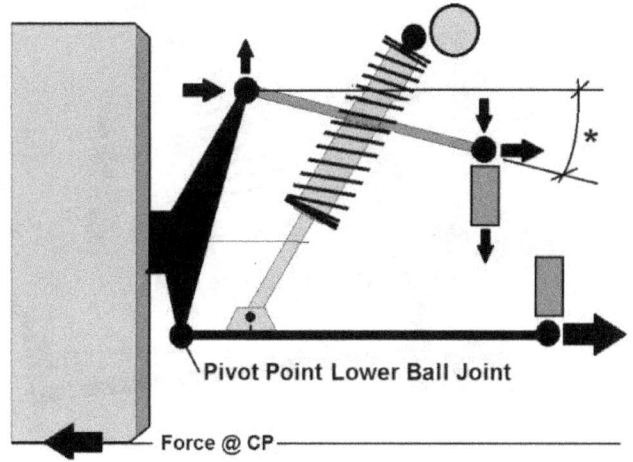

Illustration 1

Here we see the left side of a AA-arm suspension in a left turn from the drivers view. The lateral force at the Contact Patch acts to try to pivot the spindle in a clockwise direction around the lower ball joint, putting a force upon the upper ball joint. This inward lateral force pushes on the control arm and creates a vertical force being up on the BJ and down on the chassis mount. This creates an Anti-roll effect.

In the right side suspension (see Illustration 2), the lateral force is again to the right through the lower ball joint and resisted by the contact patch with that force going to the left. Again, with these control arm angles, we see an upward force at each chassis mount that together resists chassis roll.

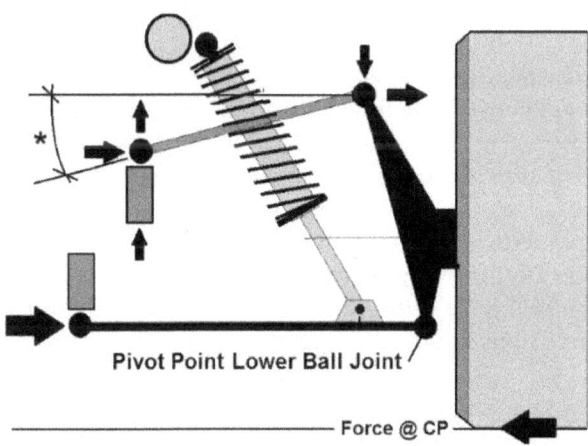

Illustration 2

This is the right side of a AA-arm suspension in a left turn from the drivers view. The lateral force at the Contact Patch acts to try to pivot the spindle around the lower ball joint in a clockwise direction, putting a force upon the upper ball joint. This outward lateral force pulls on the control arm and creates a vertical force being down on the BJ and up on the chassis mount. This creates an Anti-roll effect.

It is a similar action with the lower control arm (see Illustration 3). In this case, a lateral force equal to the force at the left side contact patch acts through the lower ball joint. If the lower control arm is mounted at an angle, then a vertical force is created that, like with the upper control arm, either increases, or reduces chassis roll. In this case, the force is upwards and adds to chassis roll.

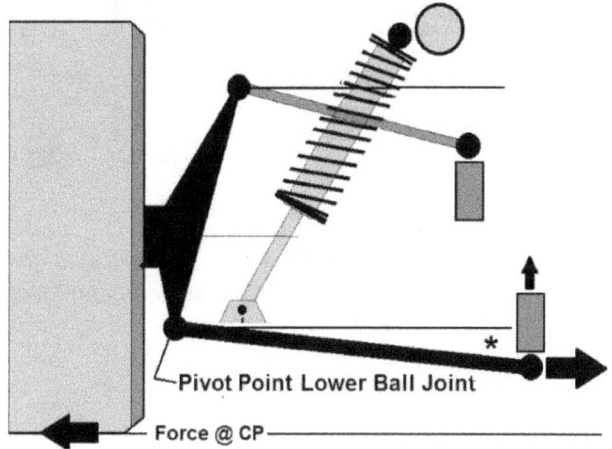

Illustration 3

Here we see the left side of a AA-arm suspension in a left turn from the drivers view. The lateral force at the lower chassis mount acts to try to pull on the lower control arm and is resisted by the lower ball joint. This then puts an upward vertical force upon the lower chassis mount. This creates a Pro-roll effect.

At the right side showing the lower control arm angle (see Illustration 4), the force on the lower ball joint is to the right and the resulting vertical force on the chassis is down, or adding to chassis roll as was the case with the left lower control arm.

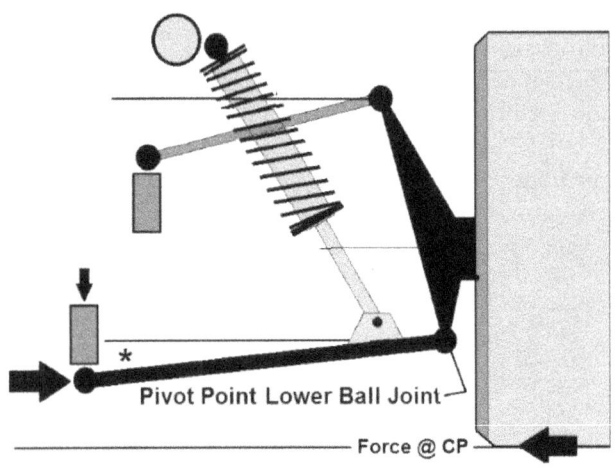

Illustration 4

Here we see the right side of a AA-arm suspension in a left turn from the drivers view. The lateral force at the lower chassis mount acts to try to push on the lower control arm and is resisted by the lower ball joint. This then puts a downward vertical force upon the lower chassis mount. This creates a Pro-roll effect.

Un-Sprung Mass Initiated JF - The other somewhat less significant JF is produced by the lateral G-force acting on the un-sprung components of the AA-arm suspension. The mass, or weight, of the wheels, tires, spindle, brakes, etc. are in themselves producing a lateral force as a result of the G-force going through the turns and therefore put a force on the control arms separate and apart from the forces applied as a result of the contact patch forces.

In Illustration 5, we can see where the lateral force created by the G-force times the un-sprung mass weight is then creating lateral forces onto the ball joints. The force for both the upper and lower ball joints are in the same direction, similar to the contact patch JF scenario except that the center of the originating force is concentrated at the Un-sprung Center of Gravity and not the contact patch or the lower chassis mount.

These forces are basically pushing on the control arms and as such are trying to push up on the ball joint end and down on the chassis mount end. We calculate the actual vertical force by multiplying the lateral force times the Tangent value of the arm angle. The calculated force is the same at the ball joint as it is at the chassis, only in opposite directions.

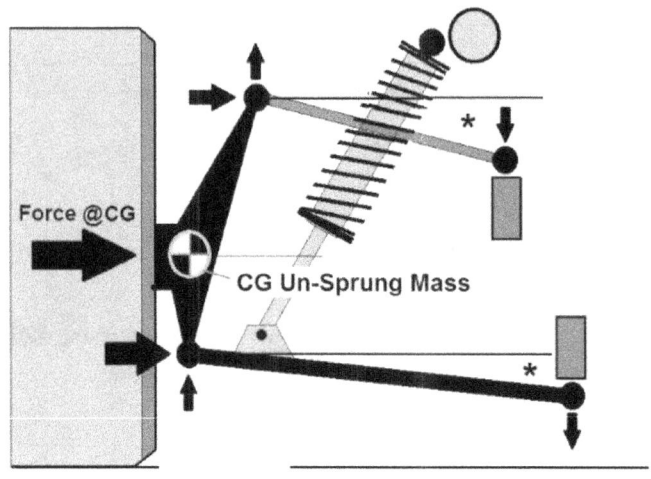

Illustration 5

In the case of the Un-sprung Initiated JF, in this view of a AA-arm suspension in a left turn from the drivers view, we concentrate on different forces. The lateral force at the Center of Gravity of the un-spring mass acts to push on the control arms. In this example, this then puts downward vertical forces upon the upper and lower chassis mounts. The net result creates an Anti-roll effect.

Then, with this layout of forces, the amount of the lateral force for each ball joint is divided between the two ball joints and are proportional to the distance each is to the Un-sprung CG. More of the proportion of lateral Un-sprung initiated force will be upon the lower ball joint and less on the upper ball joint because the upper BJ is farther away from the center of force, or Un-sprung CG.

In Illustration 6, we see the right side suspension and how the Un-sprung forces act on the control arms. These forces are basically pulling on the control arms and as such are trying to push down on the ball joint end and up on the chassis mount end. We calculate the actual vertical force by multiplying the lateral force times the Tangent value of the arm angle. The calculated force is the same at the ball joint as it is at the chassis, only in opposite directions.

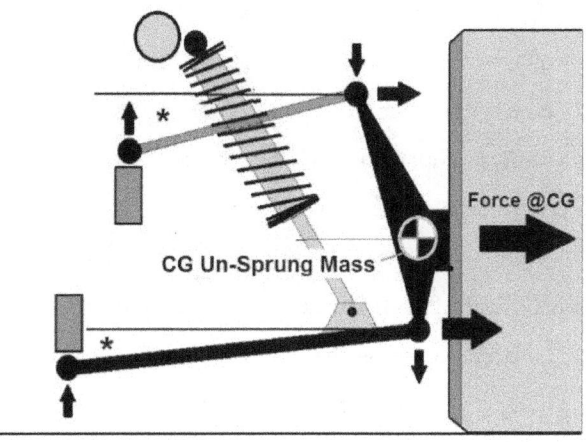

Illustration 6

In this view of a AA-arm suspension in a left turn from the drivers view, we concentrate on the right side Un-sprung Initiated JF forces. The lateral force at the Center of Gravity of the un-spring mass acts to pull on the control arms. In this example, this then puts an upward vertical forces upon the upper and lower chassis mounts. The net result creates an Anti-roll effect.

Combining Jacking Forces - To arrive at a net jacking force for each wheel, we add all of the JF forces, and translate them out to the wheel as a positive or negative wheel load that is then added or subtracted from the normal sprung mass weight transfer. The number representing weight transfer in our example being 436 would be smaller for the left side suspension and also smaller for the right side suspension.

Based on the new weight transfer numbers that now include the JF weights, our new roll angle will be less because "less weight has transferred". In reality, the same amount of weight transfers, but the weight numbers that we use to calculate roll angle have in fact changed. The JF has restricted chassis roll, but not the actual weight transfer.

This sounds complicated, and the actual computations are a little involved. We won't be going into how to exactly calculate the JF numbers, but we have explained here how ending up with certain control arm angles can have an effect on the amount of JF our chassis will have.

Chassis Roll Changes Arm Angles – We cannot use the static geometry for the suspension to analyze the Jacking Forces because as the chassis rolls, in reaction to the overturning moment dynamics, the control arm angles will change. It is important to know that as these arm angles change, so do the Jacking Force, they have to.

So, there is some back and forth calculating that needs to be done in order to arrive at accurate numbers for JF. You may have seen illustrations showing force lines acting on static control arm angles. You now know that any analysis of Jacking Force must be applied to dynamic control arm angles, and that is control arm angles as they end up at after the chassis rolls.

Jacking Force Is Dependent – One very important thing to remember is that JF is entirely dependent on how much lateral force there is on the contact patch. That is dependent on the lateral G-forces, track banking angle, tire loading, and contact patch size and weight distribution.

The Jacking Forces will change as the above described conditions change, so JF is very dynamics and moving all of the time. We can only approximate the JF for a particular setup, geometry and track layout.

The more consistent we can make the setup and the more balanced we have between the two suspension systems, the more reliable our estimates of JF magnitudes. And that is our overall primary goal with chassis setup. Just know that JF calculations is a moving target. If we know where the target is moving to, we can more easily hit the target.

Modern Applications Of Jacking Force – It seems that many teams today desire high JF numbers in the Anti-roll direction. This reduces the chassis roll and that might be a desired effect. Since the dynamic lower arm angles for most dedicated production race cars end up at a high value, a lot of Pro-roll is created and opposes the desired Anti-roll effect of JF.

What we see now is a trend to reduce the dynamic lower control arm angles so that there is less Pro-roll effect in the suspension that might counter the JF Anti-roll teams desire. The decision to use JF or not is a personal choice for race teams. We are here to present knowhow and not to influence a race teams thinking about which way to go. Hopefully our students will now have more information that will help them make critical decisions about Jacking Force.

Exam - In The Context Of This Lesson:

Jacking Force Does What?
1) Causes chassis roll
2) Provides Anti-roll properties
3) Provides Pro-roll properties
4) Adds weight to the inside front tire
5) 2 and 3

The Amount of Weight Transfer Does Not Depend On?
1) The Jacking Force amounts
2) The Moment Arm length
3) The lateral G-force
4) The sprung weight
5) The track width

Jacking Force Has Been Around For How Long?
1) Forever
2) Just the past decade
3) Since race cars became popular
4) Never before

Jacking Force Is An Affect As Long As?
1) There is lateral force
2) The control arm angles are not level
3) The car has a sprung suspension
4) All of the above

The Lateral Force At The Contact Patch Is?
1) Equal to the G-force times the tire loading
2) Equal to the G-force at the Center of Gravity
3) The sum of all of the Jacking Forces
4) G-force times the Gravity force

The Contact Patch Initiated JF Uses Which?
1) The lateral force on the un-sprung mass
2) The lateral force on the sprung mass
3) The force at the contact patch
4) 2 and 3

The Un-sprung Mass Initiated JF Uses Which?
1) The lateral force on the un-sprung mass
2) The lateral force on the sprung mass
3) The force at the contact patch
4) 2 and 3

To Determine Accurate JF Numbers, We Don't…
1) Use the static control arm angles
2) Use the dynamic control arm angles
3) Use dynamic wheel loads
4) Combine all of the JF affects

In Modern Racing, Teams Desire The Following…
1) Anti-roll effects of JF
2) Pro-roll effects of JF
3) More chassis roll
4) Less upper control arm angles

Race Car Technology – Level Three
Lesson Eleven – Rear Steer Concepts & Uses

We have talked about front geometry (Lesson 9) and we have talked about Angle of Attack (Lesson 8). Now we need to talk about Rear Steer and rear geometry. There are two primary purposes for rear steer in an asphalt race car and an additional purpose for dirt cars.

The first primary purpose for Rear Steer is that

1) it helps develop and tune the AOA, or angle of attack, for the rear suspension mostly in a solid axle rear suspension, but also in a AA-arm rear suspension to some extent. The second purpose is

2) it can help develop additional AOA for drive off the corners independent of, and in addition to, the mid-turn rear steer. The third purpose, and this is mostly for dirt cars, is

3) to develop a rotation of the car body in relation to the direction the car is traveling to produce flat plate aero side force. This is somewhat useful for some asphalt cars too to a much lesser degree.

The dirt late model cars usually have four-link rear suspensions that are setup to steer quite a bit to the right. This pushes the left side of the rear end up towards the driver and steers the rear of the car out to the side by a foot or more. The benefit is in the aero package. The flat right side of the car creates flat plate aero sideforce helping to push the car towards the inside of the turns. The angle of the rear end compared to the alignment of the front wheels is not that much different. But the alignment of the contact patches side to side is miss-aligned by quite a bit.

This Florida modified car used rear steer under power to enhance the grip off the corners. These cars are limited to an 8 inch tire and need all of the added grip they can get for bite off the corners. The mid-turn rear steer was near zero.

Why Don't Cars Have Rear Steer Instead Of Front Steer? – Steering from the rear is very difficult because the effect is so pronounced. A very small amount of rear steer goes a long way. That is why we don't see any production or race cars that steer from the rear, and that makes sense.

If, as we learned in Lesson 8 dealing with Angle of Attack, we only steered with the rear wheels, then how would the front tires develop their AOA? We know they need to be steered quite a bit to attain an AOA, so the rear would also need to be steered quite a bit just to give the front tires their needed AOA.

Once we had steered the rear wheels a sufficient amount to give the front tires their AOA, the rear would be steered way too much to the outside of the turns to ever develop the AOA they need. It just doesn't work. So, we have front steered cars.

What Is Rear Steer? - Rear Steer in a race car is an effect that is mostly caused by suspension movement. We don't have a steering mechanism the driver can use to steer the rear. Under the right conditions in measured amounts, RS can be beneficial and enhance performance. Under the wrong conditions, it can ruin your handling. We need to have a solid understanding of what produces RS and what effect RS has on the handling in our cars in order to have the best performance. And we need to know how much to use.

How Rear Steer Works - When we have rear suspension movement, as the rear suspension of the car moves, along with the controlling arms that locate the rear end fore and aft, the components for each side can move the wheel either forward or to the rear. We will use, for example, a common three link rear suspension.

If the trailing arms for a common three link are mounted at an angle from a top view where the chassis mounts are closer together, or closer to the centerline of the chassis, than the rear end mounts, then as the rear end moves laterally, there can be rear steer. This would mostly come from the movement of the panhard bar, which moves the rear end laterally as the bars chassis mount moves through an arc with the rear end mount as the radius.

In most circle track race cars turning left, the panhard bar is mounted to the chassis on the right side of the car, or to the outside of the turn. In a left turn, if the chassis mount is higher than the rear end mount, then as the right side of the chassis moves down in dive and roll, the panhard bar will push the rear end to the left until is gets to an angle that is parallel to the racing surface. This steers the rear end and rear tires so that they are pointed to the right of the chassis centerline. This tends to loosen the rear of the car.

As the chassis mount moves farther, the chassis mount is now moving below parallel to the racing surface and will now pull the rear end to the right back towards where it started out. If we know the amount of right side chassis movement vertically, we can simulate the movement of the rear end laterally.

If the panhard bar were mounted level to the racing surface, or with the chassis mount lower than the rear end mount, then the rear end would be pulled to the right with the normal chassis movement in dive and roll. This would then steer the rear end and rear tires in a direction to the left in relation to the centerline of the chassis and that would tighten the car.

What Does Rear Steer Do? - A rear end that is steered to be pointed to the left of centerline will cause the thrust angle and AOA to be left of centerline and make the car tighter on entry, tighter in the middle of the turn, and tighter on exit under acceleration. A rear end that is steered to the right of centerline causes the thrust angle and AOA to be pointed to the right of centerline and will generally make the car looser on entry, looser in the middle, and looser on exit under acceleration.

The asphalt racing surface provides a lot of traction, even on those flat "slick" tracks. Because there is very little slip of the tires on asphalt, the range of useable rear steer is very small. There are six predominant rear suspension systems used in stock car racing and all of them can produce some amount of rear steer. They are:

The Three Link System – This system has two trailing arms mounted near the rear tires and one third link, usually mounted atop the rear differential. The third link controls rear end wrap-up. The trailing arms can be mounted parallel to the centerline of the car or angled with the front mounts closer to centerline. They can also be mounted at different angles from a side view.

RS (rear steer) in this system is caused by chassis movement which can produce several secondary effects. Usually, the right rear corner of the chassis moves more than the left side, and on most flat to medium banked tracks, the left rear moves very little.

On most asphalt 3-link cars, the RR trailing arm mostly controls RS due to body roll. We usually need to position the angle of the trailing arm so that the front mount is higher than the rear mount by roughly one-third the distance that the front mount will move down during cornering. This is a generalization though and we will get more into control arm angles later on in this discussion.

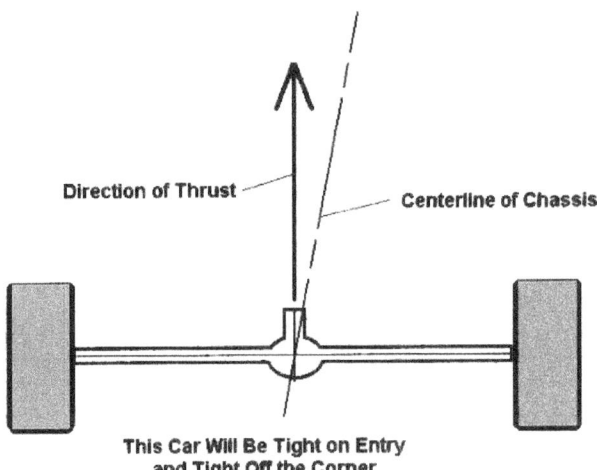

With the rear end pointed to the left of centerline, the steering will cause the rear of the car to want to run left of the front end causing a very tight condition, especially under acceleration. With the rear end pointed to the right of centerline, the car will be freed up going in, through the middle and possibly loose off the corners with the rear end wanting to run around to the right of the front. Rear steer can be tolerated in very small amounts.

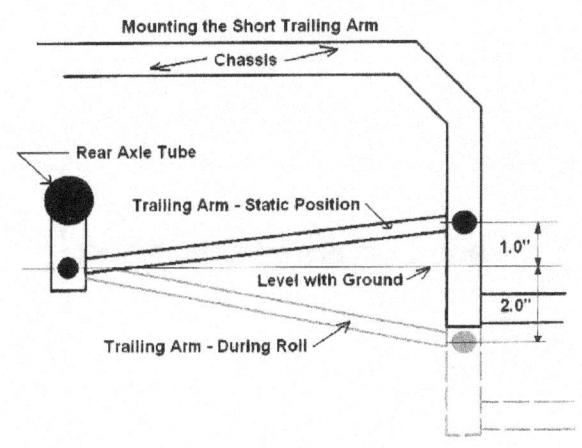

The three-link rear suspension system can produce rear steer in both directions. As the chassis moves down on the right side, the RR wheel will be moved back as the front mount approaches the height of the rear mounting point. As the front mounting point continues to move down, the RR wheel will be pulled forward. Normally, to produce a small amount of rear steer to the left on asphalt, we mount the front pivot point 1/3 of the total travel distance higher than the rear pivot.

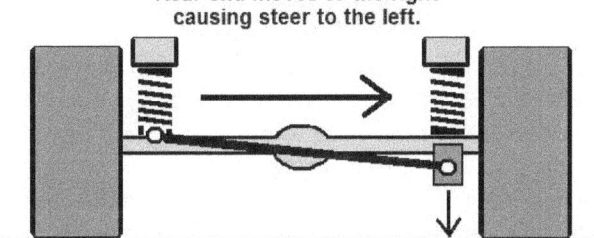

In this illustration, if the panhard bar chassis mount (to the right side) is mounted lower than the rear end mount (left side mount), then when the chassis dives and rolls moving the right chassis mount down, then the rear end will be pulled to the right. If the trailing links are angled from a top view with the forward mounts narrower, then this movement will steer the rear end to the left of centerline. At the same time, if the right link were angled so that the front chassis mount was higher than the rear mount, with the same chassis dive and roll, the right side of the rear end will be pushed back creating rear steer to the right. If we setup the movement to cancel each other out, we will end up with Net Zero rear steer. Tuning the angles of the panhard bar and the right link will allow us to fine tune the rear steer.

The Truck Arm System – The truck arm system has been adapted from the design for a 1964 Chevy pickup truck and is used on many Late Model Stock cars as well as the three premier divisions of Nascar, the Camping World Trucks, and the Xfinity and Monster Energy Cup cars. These systems react to dive and roll in a similar way to the three link suspension we described earlier. These systems can roll either to the left or to the right. The roll of the chassis and the movement of the panhard bar are the two components that influence the amount of steer in these systems.

As far as geometry related to rear steer is concerned, this can be a good system for asphalt cars. The amount of rear steer is determined by two things, 1) steer due to body roll is regulated by the height of the front mounts of the arms which are always mounted lower than the rear point of rotation which is the axle. Rear steer due to the panhard bar vertical movement are regulated by the angle of the bar, similar to the three link system.

The truck arm rear suspension system is still being used as of the date of this publication with Cup cars of NASCAR. It is a very strong system, but one that lacks proper adjustability for geometry and dynamics.

The Standard 4-Bar System – The 4-bar suspension is highly adjustable and can be made to steer in both directions. The rule about never steering the rear end to the right for an asphalt car does not apply to a dirt car. There are times when we definitely want the rear to steer to the right.

To determine the amount and direction of rear steer in this system, we must know something about the vertical movement of the corners of the car. On dirt cars, this movement can be very different than what we see on asphalt cars.

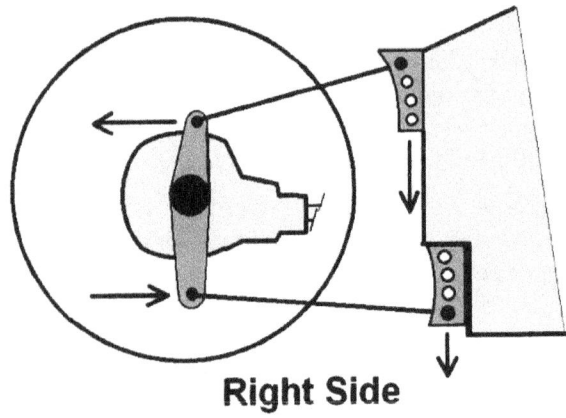

Right Side

If the bars on a 4-bar car are set in the correct holes, the movement of the top and bottom mounts at the rear end will compensate fore and aft resulting in zero rear steer. This design is best for tight, tacky tracks where we do not need rear steer and we need for the thrust to be pointed straight ahead.

As previously stated, in most cases for dirt cars, a high degree of rear steer to the right causes the body of the car to run at a high angle to the direction of the path of the car. This enhances flat plate aero forces on the flat sides of the car to push the car towards the inside of the turns when going through the turns.

The Metric 4-Link System – The metric four link is a widely used system that comes with some models of stock automobiles and is mandated for use in the stock divisions for team using those cars where it is installed from the factory. It uses four links, two on top and two on the bottom, as the name implies that are not parallel to the centerline of the car. The top links are angled from a top view with the front chassis pivots much wider than the rear pivots. The lower links are angled from a top view with the front chassis pivots mounted much narrower than the rear pivots.

With this system, the rear end stays located laterally by virtue of the opposing angles of the upper and lower links. There may be rear steer inherent in this system depending on the ride heights because the original car was designed for stock ride heights. When we race a stock car, we usually change the ride heights. Under most current rules, there is no adjustment for amounts of rear steer with these systems.

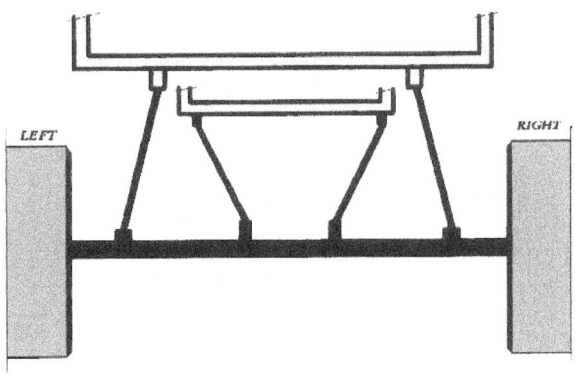

The Metric 4-link suspension has two links above the rear end and two links below the rear end. They are angled from a top view to prevent the rear end from moving side to side as the chassis experiences lateral loading through the turns.

Leaf Springs Systems – The leaf spring rear suspension system locates the rear end fore and aft as well as laterally using the leaf springs. There can be a small amount of rear steer as the chassis rolls and squats, but it is both minimal and mostly fixed as far as adjustability. The advantage of this system under certain conditions is that it keeps the rear end squared up and the thrust under acceleration straight ahead. It is useful if that is what is needed for a particular form of racing on certain race tracks.

The height of the front eye of the leaf spring in relation to the axle connection height determines the amount of rear steer. As the leaf flattens out from chassis travel, the distance between the axle and the chassis mount increases causing rear steer to the right. It pushes the right wheel back. Spacing the leaf different heights from the axle tube does not affect this action, and so the leaf spring system is very hard to adjust for rear steer.

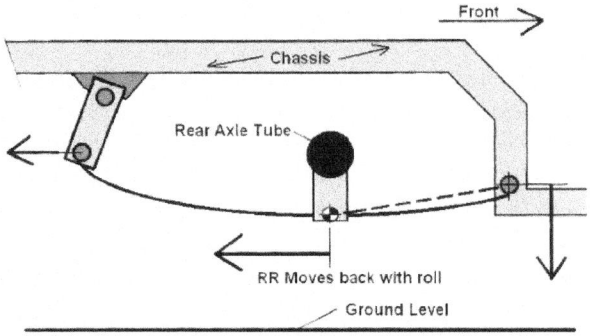

In the leaf spring system, as the front leaf mount moves down, the spring flattens out and moves the axle back. This creates rear steer to the right. Because the rear mount for the leaf spring is on a shackle, its motion will not affect the fore and aft location of the axle. Rear steer is hard to adjust for a leaf spring system.

The Z-link System – The Z-link rear suspension, or swing arm as it is sometimes called, is another system used mostly on dirt cars, but has been used on asphalt cars as well. Compared to the 4-bar cars, it has more limited adjustment for rear steer and historically has worked well on the tighter and more highly banked race tracks because the direction the rear end is pointed is more straight ahead.

Some manufacturers have added multiple mounting points on the front and rear chassis mounts on these systems to make them more adjustable like the four-link systems. This helps make the rear steer characteristics more adjustable for changing conditions, but can never make them steer like the four-link suspensions.

A Z-link suspension system uses a link extending from the under the rear axle tube forward to the chassis and one from the top of the rear axle tube rearward to a mount on the chassis. Most designs use very few mounting holes that would enable the team to adjust for the amount of rear steer in these cars. This one has quite a few. The Z-link was originally designed for reduced rear steer.

Measuring Your Rear Steer – To measure the rear steer in your car, first know the shock travels at the four corners of the car. Then you can recreate the rear attitude at mid-turn. You can use a tape measure for simple measurements or a laser system for more detailed analysis. With the tape, simply measure, at ride height, from the rear wheels towards the front of the car level with the ground to a point. This can be a mark on a tape near the front wheel, or the actual edge of the wheel well. For the laser systems, follow the manufacturer's instructions.

Move the rear wheels in any increment you choose to the position they are when the car is at mid-turn and then re-measure the distances. If the movement for the left and right wheels are in opposite directions, add them to determine the amount of rear steer. If the movements are in the same direction, subtract them. You might be surprised at how much rear steer is going on.

With asphalt circle track cars, we are finding ways to minimize rear steer on entry and through the middle of the turns, and then creating rear steer to the left to enhance bite off the corners through other means.

Creating Zero Rear Steer – Because in some systems the panhard bar causes rear steer as well as the rear link angles also cause rear steer, we can adjust the angles of those two in order to create offsetting steering for a Net Zero rear steer. In my many years in racing, I have never considered this concept until recently. Better late than never.

Suppose we can determine that the panhard bar, with an angle caused by a higher chassis mount than the rear end mount in a three link suspension for example, causes rear steer to the right. In this same system, our right trailing link, by having the front mounted level to, or lower than, the rear mount, causes rear steer to the left. If I adjust those two, the panhard bar angle and the link angle, to create equal but opposite steering angles, I will end up with NZ rear steer. And maybe that is what I want.

Or, maybe I want a combination of the two movements that will lessen the rear steer that might occur from just one of the two movements. I can then fine tune my rear steer if I know how the two movements are working.

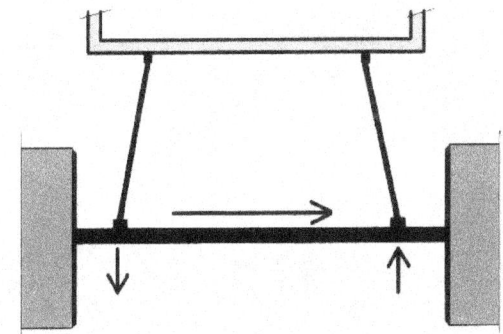

Most three link rear suspensions are designed with the forward mounting points for the outer links to be narrower, or closer to centerline, than the rear mounts attached to the rear axle tubes. With this design, as the rear end moves laterally, the rear end steers. If the rear end moves to the right, the wheels steer to the left.. The left link causes a greater distance from the axle tube to the chassis and the right link causes a lesser distance from the axle tube to the chassis. This motion steers the rear end and wheels to the left. The static panhard bar angle determines which way the rear end moves and which way it will be steered.

Rear Steer Under Power – Because the rear tires are providing all of their grip to keeping the rear of the car on the track through the middle of the turns, when the car begins to accelerate off the corner, the rear tires need more grip to handle the extra work. There are several ways to do that in keeping with the idea of not changing the middle handling characteristics.

Way Number One – The first way to create rear steer under power only is to use a pull bar, or lift arm. This causes the rear end to rotate when power is applied. The rotation is a product of the pinion gear trying to climb the ring gear in the rear end. This put a force into the rear end trying to rotate it clockwise from a left side side-view.

So, the rear end will rotate some amount when using a pull bar or lift arm. If we stagger the height of the trailing arms on a three link rear suspension, then when the rear end rotates, the left axle hub will move rearward more than the right axle hub. This produces rear steer to the left and adding Angle Of Attack to the rear tires. This provides more grip for the rear tires to counter the lateral forces and the accelerating forces.

The amount of rear steer is limited, but we don't need, or can use, much rear steer. The average amount the right rear tire moves ahead of the left rear tire is in the range of 0.050" to 0.075". This doesn't sound like much, but it is sufficient to do the job. Any more than this and the car becomes too tight off the corners.

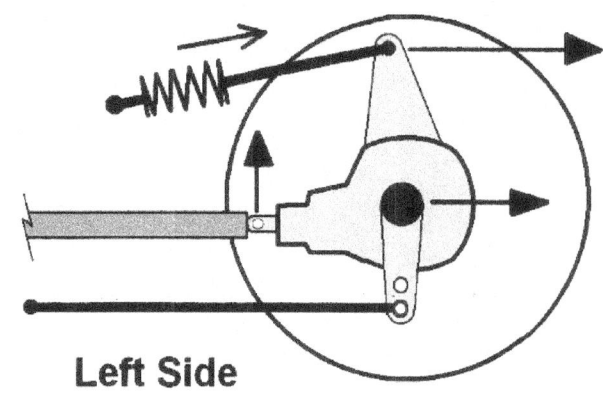

Left Side

As described above, we mount the left side link in the bottom holes on the bracket coming off the rear end axle tube on the left side.

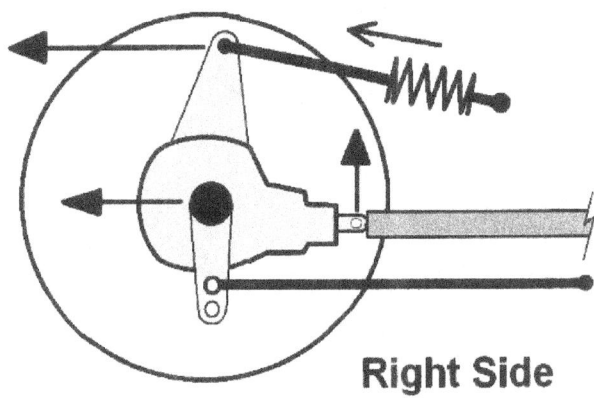

Right Side

On the right side, we mount the link in the top hole of the bracket coming off the axle tube. The distance that the right side moves rearward comparted to the left side is less. This steers the rear end to the right of the chassis centerline creating rear steer to the left.

When using a pull bar third link such as this one, as the car accelerates, the bar is pulled back due to the engine torque trying to rotate the rear end. With this movement, the rear end moves back. To produce rear steer, we stagger the heights of the trailing links, the right side higher than the left side links. With this staggered height, the left side moves back more than the right side creating rear steer to the left. The actual difference in movement between the left and right wheels is in the range of 0.050" to 0.075". This steer adds to the AOA of the rear tires and helps provide more rear grip.

Way Number Two – Another way to produce rear steer only under acceleration is to install a push-link or compressible right rear link in a three link suspension. This can also be done with the right side links on a Metric four link suspension.

As the rear tires drive forward under acceleration, a high amount of pushing force is transmitted into the suspension links. If the right side link, or links, are compressible, or able to compress, then the link becomes shorter and the right wheel moves forward and the rear end steers to the left providing more AOA.

The amount of movement of the right wheel forward using this method must be regulated so that it does not become excessive. With the Way Number One method, we see where we only need an average of 1/16th of an inch of movement to do the job. If your compressible link moves ¼ inch or more, obviously that is too much in most cases.

This is called by several names including push rod, spear rod, or just compressible link. As the right rear tire pushes forward during acceleration, the rubber biscuit compresses and the link shortens. This movement steers the rear end left of centerline and creates more AOA for the rear tires. The amount we allow this link to move is critical to how much effect it will have. It can be overdone by far. Movement of 0.250" to 0.375" is not uncommon and that is entirely too much in most cases. The rubber can be pre-compressed so that the movement is reduced.

The beauty of these two methods is that the rear steer only occurs under acceleration and does not interfere with the balance of the setup through the middle of the corner. And, these two can be tuned so that you get the exact amount of rear steer you need for your car at your race track.

Conclusion – Rear steer is an important concept to understand and apply and in reality, one that can either enhance our setups, or ruin an otherwise great balanced setup. It's use and magnitude is specific to the type of racing you will be doing, but to ignore this mechanical effect means you are tossing the dice on how your car will perform on the race track.

Rear steer is one of those chassis setup areas that needs to be addressed, evaluated closely and applied in measured amounts. Through careful analysis and planning, rear steer under power can enhance the overall performance of any race car.

Exam - In The Context Of This Lesson:

The Primary Purpose for Rear Steer Is To?
1) Loosen the car
2) Tighten the car
3) Create additional angle of attack
4) Make it easier for the driver

Virtually All Cars Driver Steer From The Front Because?
1) It is too expensive to design rear steering cars
2) The technology isn't well developed
3) It is closer to the steering wheel
4) The AOA geometry is all wrong when steering from the back

Rear Steer Can Be Caused By?
1) Chassis dive
2) Chassis roll
3) Acceleration forces
4) All of the above

Is Zero Net Rear Steer Possible?
1) Yes
2) No

We Can Regulate The Degree Of Rear Steer By?
1) Changing link angles
2) Changing panhard bar angles
3) The use of compressible links
4) All of the above

Acceptable Levels Of Rear Steer Are Measured As?
1) Ten thousandths of an inch
2) Hundred thousandths of an inch
3) Inches and fractions of inches
4) Degrees and decimals of a degree

The Least Adjustable Rear Suspension For Rear Steer Is?
1) The three link system
2) The four-link system
3) The truck arm system
4) The leaf spring system

Rear Steer Can Be Used To?
1) Provide more AOA
2) Provide less AOA
3) Help create flat plate aero for added side force
4) All of the above

Race Car Technology – Level Three
Lesson Twelve – What Makes Traction? A Review

Here we want to review the three major contributors to making traction. This is a combination of Lessons 2, Lesson 6 and Lesson 8. We want to emphasis the basic methods we use to gain grip, or traction. Once we have these three methods firmly in our grasp and fully understood, we can proceed to working out our setups.

Race car performance is all about gaining speed in the slowest portion of the race track and then carrying that speed all of the way around the race track for a faster overall average speed. That is what makes up a lower lap time.

A race car that has all of the characteristics of what makes traction will carry a lot of speed through the turns and this will make the speed around the entire lap faster. In this Lesson we will review What Makes Traction.

Load Distribution On The Four Tires – In almost every situation for every type of race car, we will never have equally loaded inside and outside tires. This is due to the concept of weight transfer during cornering. Even with a biased left side weight common in circle track racing, we will never be able, under the current rules sections, to have enough left side, or inside, weight to overcome the weight transfer amount. So, we will never have dynamically equally loaded left and right side pairs of tires.

If you remember in Lessons 4 and 6, we discussed how load transfers during cornering and how we needed to end up with equal loading on the outside tires, and equal loading on the inside tires in order to get the most traction out of each pair of tires at each end of the car.

Once we have developed a balanced setup where we can predict and plan for the ideal weight transfer and load distribution on the four tires, we will then have maximum grip from those four tires. But it doesn't stop there. There are two more concepts of making traction that we need to address and perfect.

In short, equally loaded tires provide the most traction from an opposing pair of tires. Equal un-equal loading provides the most traction for a set of four tires on a race car. If we have equal amounts of un-equal loading on the pairs of tires on the two axles, then we have the most overall race car grip possible from a tire loading standpoint.

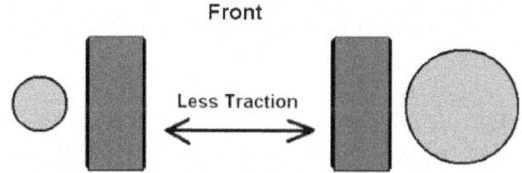

The result of an unbalanced setup where excess load transfers to the Right Front tire causing the rear tires to be more equally loaded. This would be a very tight handling car.

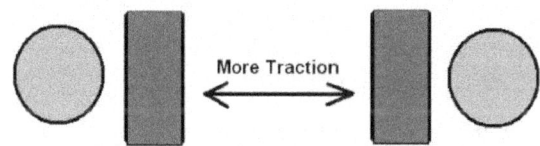

Two tires on the same "axle", or same end of the car, will have the most traction when they are equally loaded. We covered this before explaining that even though two tires have the same total load, when they are un-equally loaded, they have less grip as a pair. On this car, the rear has more equally loaded tires and therefore more grip than the less equally loaded front tires. This car will be tight and not want to turn very well.

Ideal Dynamic Weight Distribution For the Low Static Cross Weight Range

With a few changes to the setup we can create a setup where the tire pairs at each end are what we call equally un-equally loaded and therefore have the same grip level. This car is neutral in dynamic tire loading and neutral in handling.

Contact Patch Size - In the Lesson 20 in RCT Level Two school, we discussed how having the right camber can increase the tire contact patch size. If we have maximized the loading on the four tires, then if we can increase the tire contact patch area, we will then have even more grip from that tire.

The discussion mostly involves maximizing the tire contact patches on the AA-arm suspensions. The straight axle suspension is hard to work with and limited on what you can do to improve the cambers and contact patch for those tires.

So, we work to get a proper load distribution on the four tires, and now we work to get the maximum tire contact patch size to better utilize that perfect loading. But we also need to understand how a tire makes lateral grip to counter the centrifugal forces trying to push us off the track.

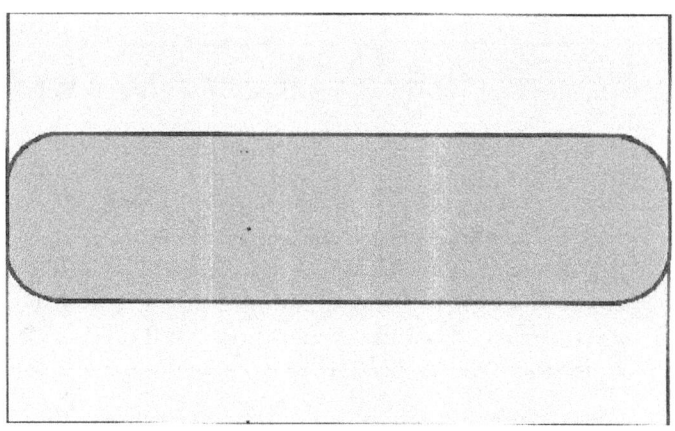

A front tire contact patch can be maximized by adjusting the camber of the tire. We want to adjust the front cambers using tire temperature and/or wear patterns to tell if the camber can be improved. We then need to maximize the pressure distribution on the contact patch to create the largest contact patch possible. This sketch shows how our contract patch pressure distribution would look if the tire temperatures were even across the face of the tire. This is not ideal for the front tires on a circle track car. It may well be perfect for a race car with wide tires with a low sidewall profile.

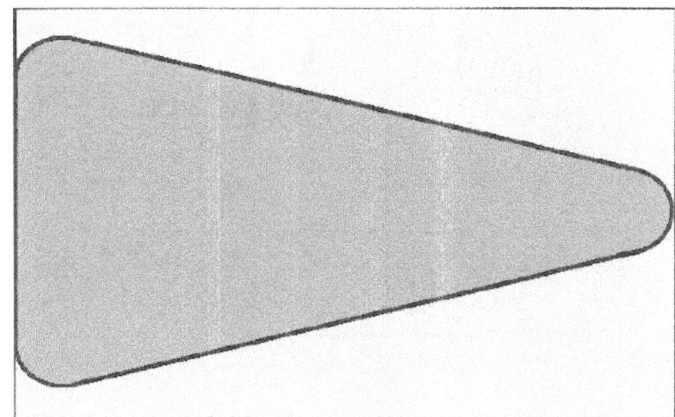

If we introduce more camber to a tire run on most circle track race cars, where the inside edge temperatures are higher by more than 15 degrees from the outside edge, and reduce the pressures, then the contract patch would look somewhat like this. The area of this contact footprint is larger than when we had less camber and uniform tire temperatures across the contact patch. Therefore, it has been proven that this cambered pattern will have more grip for those cars. For wider tires with a low sidewall profile, this may not provide the stated benefits.

Angle Of Attack – The third ingredient for making traction is creating angle of attack, or AOA. When we have the perfect loading on the tires and the maximum contact patch we can possibly achieve, we will still not have traction and lateral grip if we don't have AOA in our tires.

AOA is, as discussed in Lesson 8, how we produce Centripital force, which is the force the tires use to counteract the centrifugal forces pulling the car to the outside of the turns. Something must resist the lateral forces, or G-force caused by cornering. That something is AOA.

Both the front and rear tires must have some degree of AOA in order to negotiate a turn at high speeds. The exact amount of AOA needed depends on the speed, banking angle, tire grip levels, and track surface grip levels. Our speed is really regulated by how much lateral grip our tires can produce. The more grip the tires produce, the faster we can go.

How Tires Develop Angle Of Attack

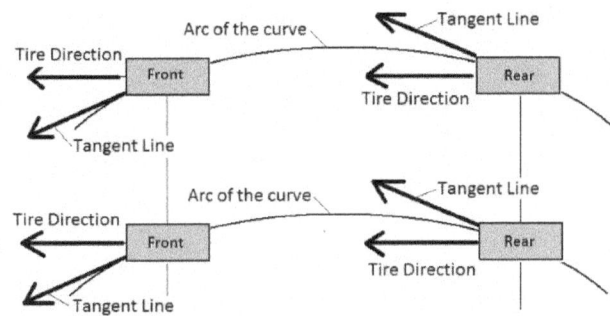

This sketch used in a previous Lesson shows how we need Angle of Attack to generate the resistance to lateral forces commonly called lateral G-forces. A tire cannot resist the G-forces if they are not steered, or pointed, to the left of the arc tangent line. The front tires must be steered to a slight angle past and inside the tangent line. The rear tires already have an AOA angle and we may need to add to, or take away from that angle to provide the tire with the proper AOA for the race track we are racing on.

Make sure you fully understand these concepts of how traction is made. If you need to go back and review the previous lessons that explain in detail these concepts, by all means do that. We will now proceed into the setup part of the Level Three of the ORS where we will plan out and evaluate our setups.

Exam - In The Context Of This Lesson:

The Ideal Dynamic Load Distribution Is?

1) When all four tires are equally loaded

2) When the front and rear pairs of tires are equally un-equally loaded

3) Never possible under the current rules packages

4) Always possible for every setup combination

For Circle Track Race Cars, The Best Contact Patch Is?

1) When the tire pressures are ideal.

2) When the heat on the inside of the tire is more than at the outside

3) When there is even tire temperatures across the face of the tire

4) When there is even wear across the face of the tire

To Provide Proper Angle of Attack, The Front Tires Are?

1) Already at some angle of attack

2) Needing to be steered parallel to the tangent line

3) Needing to be steered just inside of the tangent line

4) Needing to be steered to follow the arc

To Provide Proper Angle of Attack, The Rear Tires Need?

1) To be steered just outside of the tangent line

2) To be steered parallel to the tangent line

3) To be steered just inside of the tangent line

4) To be pointed to follow the arc

Race Car Technology – Level Three
Lesson Thirteen – Force Verses Weight

In the process of setting up and evaluating a setup, we need to understand the relationship of Force verses Weight. Once we have knowledge of how they relate and what each one is, we can then move on in our studies. This is a critical step that is necessary to go through before we move on. It will become very apparent once you begin the setup portions. This is a short presentation, but one where the concept will be referred to over and over again.

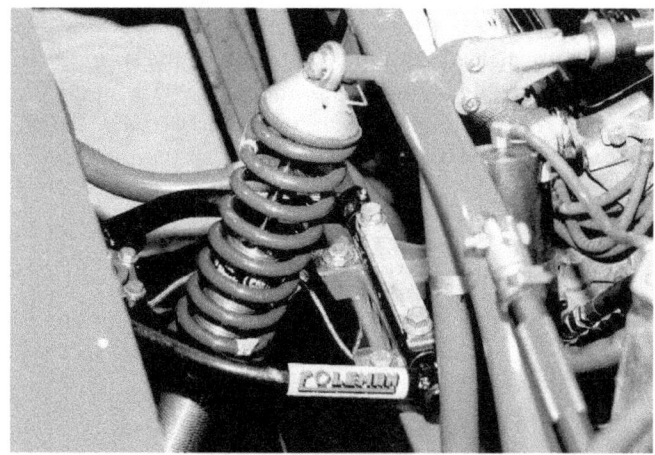

We can easily measure the length of the spring when it is installed in the car at normal ride height. We subtract the installed length from the free height length (while not under load) and multiply that difference times the spring rate in pounds per inch. This gives us the force the spring is exerting on the suspension to support that corner of the car.

Weight is easily measured and read on common racing scale pads. We weigh the car at static ride height on a level surface. This provides the gravitational loading on each tire. We can adjust the distribution of the loading easily on a race car.

Weight Defined – Weight is the measure of the force of the gravitational pull on a body and is read in pounds, or any other denomination of measurements. Everything on Earth is pulled towards the center of the earth by a force called gravity. In a race car, we can measure static weight using scales. That is, when the car is not moving, we place the car onto scales designed to provide a reading of the amount of weight that tire supports.

In a race car, going around a turn, the weights on the four tires change. They change because of 1) weight transfer, and in the case of a banked race track, 2) mechanical downforce. The mechanical downforce is added to the gravitational weight. And, mechanical downforce is the combination of gravity and the lateral force we know as centrifugal force.

If we could measure that combination of centrifugal downforce and gravitational weight that is a result of weight transfer, we would be able to determine the spring force we would need to support that combined weight. But, I'm getting ahead of myself. Let's now define Force.

Force is force and a solid link will exert the same force as a spring to hold up this corner of the car at normal ride height. If we had a load sensor installed in the solid link, we would read the same force as the spring would provide to hold this corner of the car at ride height.

Force Defined – Force is the energy needed to support weight. If the load on the right front tire on our race car weighs in at 650 pounds on the scale, it would take some amount of spring force to support that weight. For a normally setup asphalt late model, the spring force would be in the neighborhood of 985 pounds. We found that by squaring the motion ratio of wheel travel to spring travel (0.660) and dividing that number into the scale reading.

For a balanced setup, there is a load at each wheel that represents the ideal loading for the four tires at mid-turn at speed. If we can determine the spring travel that provides that loading, we can then determine the spring force being used by the suspension.

In the place of data acquisition, we can record the maximum spring travel that represents the total spring force needed to support this corner of the car. We need to make sure we don't influence this reading by heavy braking into the corners that might present a greater amount of spring travel than what occurs at mid-turn.

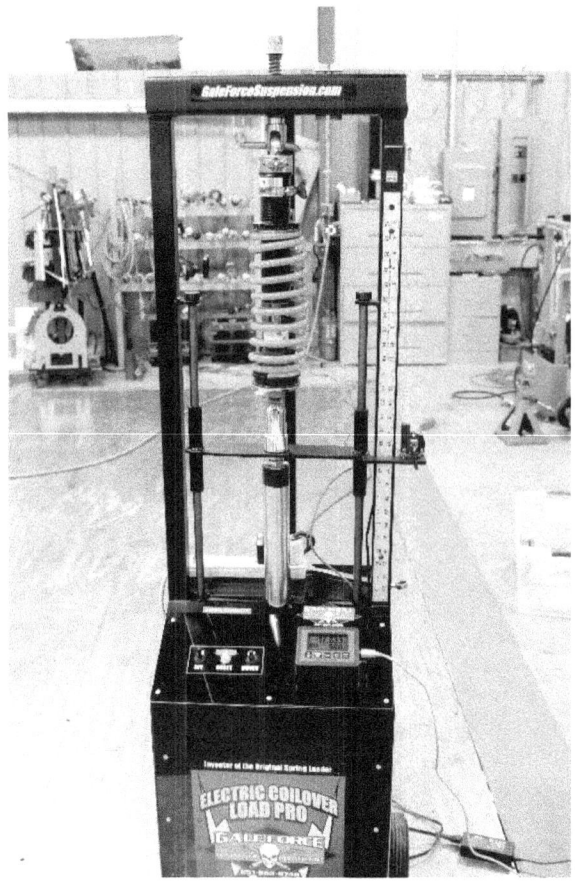

The Gale Force Load Pro measures the loading on the spring at mid-turn shock travel for a coil-over spring. By compressing the spring to the mid-turn length, the loading is then shown and compared to what we determine to be ideal loading.

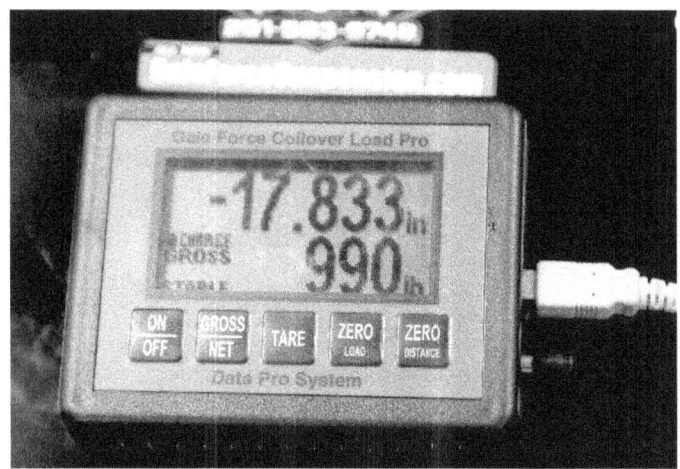

Here is the readout on the Gale Force Load Pro with a car at ride height. This number comes very close to our sample car force calculation of 985. On this machine, the team would continue to travel the spring until it reached the mid-turn length and then read the force directly. There is no other way to measure the dynamic spring force that represents the car on the track.

Calculating and Measuring Force - If we installed a spring with a rate per inch of 250 ppi (pounds per inch), then if we divide 985 by 250, we get a spring travel of about 4 inches. In fact, that is exactly how we can determine the force a spring is providing. If we measured the installed 250 ppi spring at 10 inches in length with no load on it, then if we measured it at a length of 6.0 inches on the car, holding up the RF corner, we would know the spring force was 985 pounds.

If we somehow knew, or could calculate the load in pounds that the RF tire would support at mid-turn, at speed, we could then determine what spring force would be needed to support that dynamic weight. In previous calculations using a sample race car, we came up with a RF dynamic tire load of 1575. The force needed to support that load is calculated the same way as when it was at rest. The ratio is the same.

With the static weight verses dynamic weight ratio of 0.6600 (650 / 985), if we divide the RF corner dynamic load of 1575 by that ratio, we come up with 2,386 pounds of force. The ride spring force, plus the bump force, plus the sway bar spring force would all need to add up to 2,386 pounds to support that weight.

This is the essence of measuring and evaluating a setup. If we can determine the ideal loads the tires would support that would create the ideal dynamic load distribution, we could then compare the ideal spring force to what the car was actually doing and then make adjustments to bring the setup into balance. See Lesson 4 and 5 for more information on that.

The sway bar can provide a spring force that will help support the Right Front corner of the car. The sway bar is a spring and when twisted, or otherwise put into play, adds to the overall force needed to support the cars suspension at mid-turn.

Summation – Get this concept firmly in your grasp because we will be working with weight verses force as a major part of race car setup in the remainder of Level Three. We discussed some of this in Lesson 7, but we feel the need to drive this home to the students. It is so much a major part of working out the balance of the setup as we get into the rear world of race car setup and tuning.

Exam - In The Context Of This Lesson:

Weight Or Load Is The Measure Of What?

1) Spring Force

2) Gravitational pull on a body on earth

3) The combination of gravitational pull and centrifugal lateral force

4) 2 and 3

Force Is The Measure Of What?

1) Ride spring under compression

2) Bump device under compression

3) The sway bar under torsional loading

4) All of the above

Which Provides The Force Needed To Support Dynamic Loading?

1) The ride spring

2) The bump device

3) The sway bar

4) All of the above

Lesson Fourteen – Where Does Force Come From?

Now we have defined weight verses force and now know that force is the mechanism that supports weight or loading. We need to now explain where this force comes from so we can know how to make changes to the forces and the contributors of those forces in order to setup the race car.

The knowledge of how the forces are administered and where they come from is relatively new. Individuals and companies who specialize in race car dynamics prediction and evaluation that have existed over the past five to ten years have combined their knowledge into what we are presenting in this school and in these lessons. It's not like this is old school knowledge, quite the contrary, this is extremely new school.

The force supporting the front suspension, and especially the right front, on this super late model asphalt race car is a combination of the ride spring force, the bump device force, and the sway bar force. We'll explain how all of that works in this Lesson.

Force Comes From Ride Springs – The first place the forces that hold and support the loads at the four tires comes from is the ride springs. This is fairly obvious, but important to point out and understand. Because there are different designs of race car, we need to define what the various ride springs are and how we might measure the force on those springs.

In a modern race car, the ride springs can be either coil-over springs, stock type "big" springs, torsion bar springs, and leaf springs. Each of these springs has a unique way they are mounted in the car. Let's examine each type of spring one at a time in relation to its role in creating force.

The Coil-over Spring – This is a very common type of modern spring that is found in not only pavement late models, but also dirt late models, dirt and asphalt modifieds, quarter midgets, prototype cars, and formula race cars. I might have left out a few types, but you get the picture.

The coil-over spring is mounted on a shock body most of the time. It can be mounted on a slider for some dirt car applications. It operates with a motion ratio meaning that it moves a lesser distance than the wheel, or in some cases a greater distance. It all depends on the motion ratio.

As covered before in previous lessons, the force is measured by dividing the motion ratio into the wheel weight. That is, whatever weight the wheel and tire supports, be it at static ride height or on the track at mid-turn, we can know the spring force by dividing the wheel load by the square of the motion ratio. This gives us the total force needed to support that weight or load, but it might not all be supported by the ride springs.

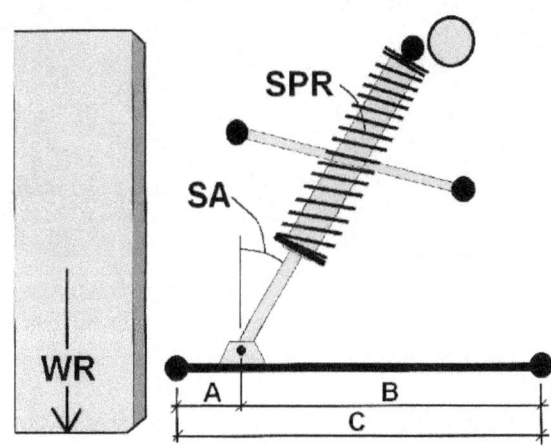

The force supplied by the ride spring is calculated using the motion ratio divided into the load the right front tire carries both statically on the scales and dynamically on the race track at mid-turn.

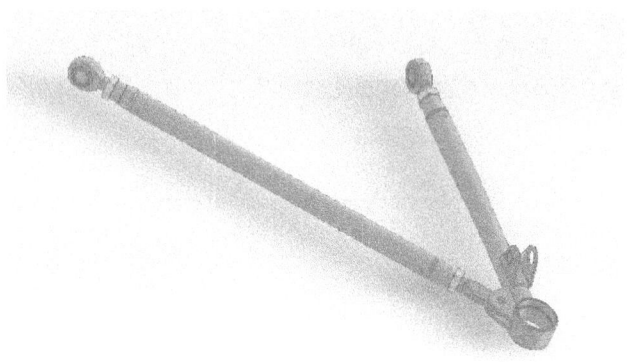

This sketch of a right side lower control arm on a super late model car shows the main link (right side), and the coil-over mount tabs. The tabs are about 2.50" from the center of the ball joint housing in red, and about 15.0" from the chassis mount for our example. The left link is the strut going back to the chassis.

This is a much better depiction of a formula car and the coil-over design. The link coming from the lower control arm (center link on the right side of the photo) is connected to the rocker (black piece to the right of the right spring). Then this pushrod rotates the rocker to push the coil-over in compression. To find the motion ratio, we remove the coil-over, place a measured thickness of block under the tire and then measure the movement of the mounts for the coil-over. If we space the wheel up 1.00" and the coil-over compressed 0.75, then our motion ratio is 0.75 / 1.00 = 0.750. To find the force to support the wheel load, we divide the motion ration squared (0.5625) into the load amount. For a 600 pound load, we need a force of (600 / 0.5625) 1,066 pounds.

Formula cars run coil-over shocks in most of the classes. The tire/wheel is connected to the rocker through a push rod (angled link shown in the right front going up to the upper part of the body). Then the rocker, inside the body, is connected to the coil-over. The motion ratio for this car is measured the same as for a late model, the movement of the coil-over divided by the vertical movement of the wheel.

The "Big" Spring – What we normally call a big spring car is one where the front, and rear in some cases, is supported by a larger spring that resembles the stock spring that comes in certain production automobiles. In stock car racing, this spring is mounted between the lower control arm and the chassis at the front of the car, and onto the rear axle or truck arms at the rear.

As for the front mount, the motion ratio is much different than we find in the coil-over application. The big spring is mounted much farther from the ball joint and therefore must be much stiffer in rate to support the same wheel weight verses the coil-over spring.

Whereas the coil-over spring might have a motion ratio of 0.850, a big spring car might see a motion ratio of only 0.645 or so. When we square those numbers, we get a value of 0.722 for the coil-over and only 0.416 for the big spring. It takes 1.7 times the coil-over spring rate for a big spring to support the same load. If we installed a 150 ppi (pound per inch) rated coil-over spring, to create the same force in a big apring, we would need a 265 pound spring. This is important for those who run the big spring cars and want to emulate the coil-over setups.

length of the torsion bar arm. A sway bar is a type of torsion bar, but we'll get more into that later on. And, the torsion bar has a motion ratio depending on where it is mounted to the lower control arm.

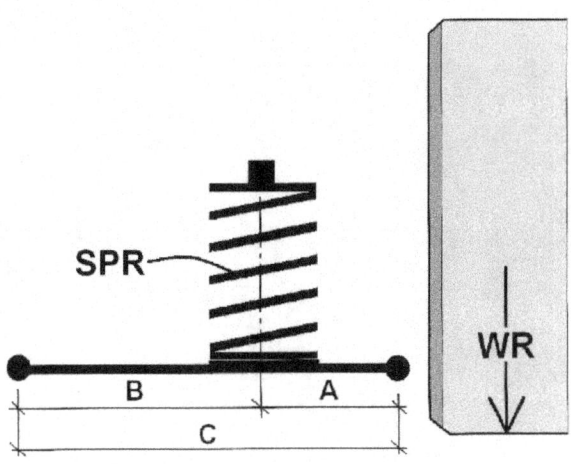

The motion ratio for a big spring car front suspension is calculated the same as a coil-over car, B divided by C, A being the distance from the ball joint to the center of the spring. Because the bump for these cars is on a much different motion ratio, we would translate this spring rate to a similar spring rate as if it were a coil-over mount. Since the bump is usually mounted on the shock mount, we would use the shock motion ratio for this calculation. Then we can just add the ride spring converted rate to the bump spring rate.

Sprint cars usually use torsion bars as the suspension springs. Some use torsion bars in the rear and coil-overs in the front. It all depends on the design of the car. The torsion spring is much like a sway bar, but does actually act like a spring and not an anti-roll device.

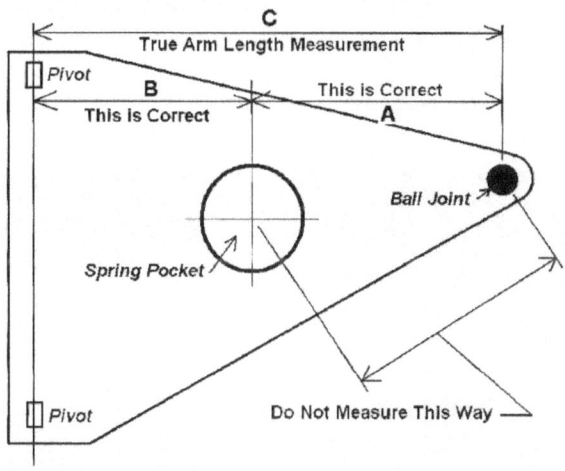

The correct way to measure the lower control arm length for a stock dimensioned lower control arm is this way. The measurement from the ball joint and center of spring to the chassis mounts is at a ninety degree angle to those mounts.

Here we can see the arm mounted to the bar that runs through a tube under the rear body under the front of the number "8". The silver arm just to the right of the left rear tire runs forward to lay on top of the rear axle tube. Above that arm is the ride height/preload adjuster for the right side torsion arm.

The Torsion Bar Springs – These types of springs are mostly used in sprint cars. Some older stock cars have used these springs because the stock automobiles used them. They are not popular in most modern race cars.

The rate is affected by the length of the bar, the diameter and wall thickness of the bar, and the arm

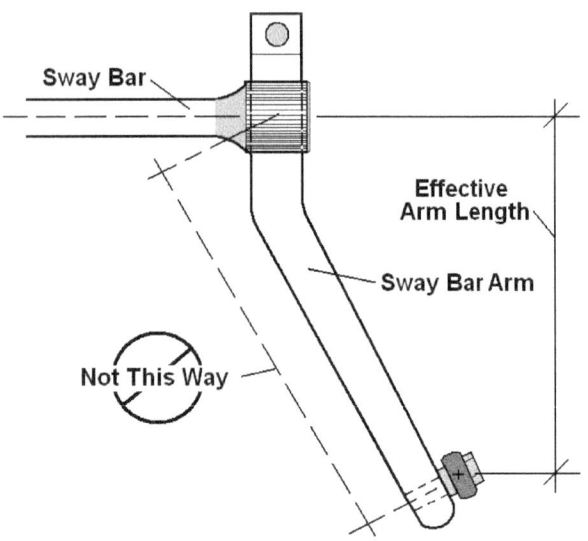

When we refer to the arm length for a sway bar or torsion bar, we are referring to the Effective Length and that is the distance at right angles to the bar of the center of the mount to the lower control arm to the centerline of the bar. The sketch shows the way to measure the arm length.

This steel coil spring bump is now in play since the shock body has traveled far enough to make contact with the bump spring. The bump spring rate is added to the ride spring rate. When we measure force, we compress the shock to the mid-turn length and then read the force. This number represents the ride spring force and the bump spring force. If we remove the ride spring and compress into only the bump spring, we can know the force generated by the bump spring only.

The Leaf Spring – The leaf spring is a type of ride spring normally used on the rear of a straight axle race car. Its rate depends on the number of leafs, the thickness of the leaf, and length of the leaf. The leaf springs are almost never used on the front of a race car, so for this discussion, we'll not include leaf springs for providing force.

Force Comes From Bump Springs - Every type of bump device is a spring. The different types react differently to the application of force. Whereas a coil bump spring and the bellows springs have linear spring rates, the bump stop type of spring has a progressive spring rate in compression, and digressive rate in rebound.

The bump spring only comes into play when the chassis has traveled far enough for the bump device to come in contact with a stop that represents the chassis. For a coil-over mounted bump, it is the shock body that contacts the bump device. When it comes into play, its rate is added to the rate of the ride spring in relation to the motion ratio of the bump to the motion ratio of the ride spring.

The wheel rate of a bump spring is dependent on the motion ratio, just like the coil-over and big springs are. If the bump is mounted on the shock shaft, then it has the same motion ratio and wheel rate ratio as the shock. If it is mounted separately from the shock, then it may have a different motion ratio and different wheel rate ratio. Remember this point.

This is a commonly used bump that is very soft and very non-linear in spring rate. It gains rate as a slow pace and then at a certain point begins to rapidly gain spring rate and force.

calculated just like a coil-over or big spring, by its mounting distance from the ball joint verses the lower control arm length.

If the motion ratios are different side to side, then the sway bar can become like a ride spring. This is good to know and we'll delve into that deeper in the next Lesson. Just know that the larger the sway bar diameter, the stiffer the spring for roll resistance as well as ride spring uses.

In determining the total force needed to support the tire loads at mid-turn, the contribution of the sway bar spring rate and its force needs to be taken into account. Again, we will delve into that more later on, but at this point, know that the sway bar spring rate can contribute to the aggregate force needed to support the tire loads.

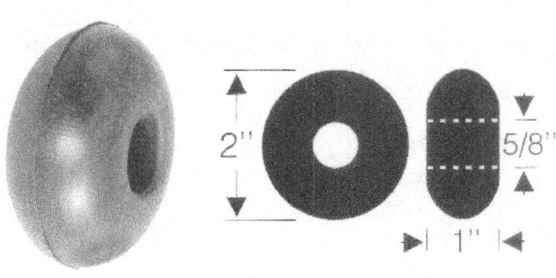

This type of bump stop is harder and provides more spring rate. These can be stacked for more travel and a softer bump spring rate. At some point of compression, these go basically solid.

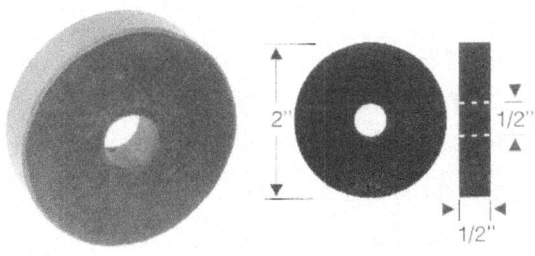

The "hockey puck" type of bump device has less travel before going solid, but provide a soft, yet stiff, spring rate. These too can be stacked for more travel and less spring rate.

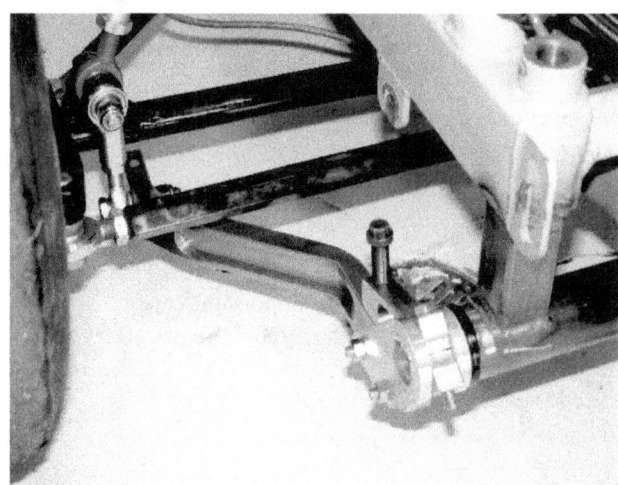

The sway bar is actually a torsion bar and provides an additional spring rate, mostly coming into play when the chassis rolls. For this installation, the sway bar has nearly the same motion ratio as the ride spring in the coil-over assembly because it is mounted to the bottom shock mount. There would be a slight difference in motions due to the shock mounting angle. We could find a point along the lower control arm where the motion ratios are then exactly the same. That point would be a short distance towards the chassis mount of the lower control arm from where it is now.

This illustration shows how bumps are mounted on the shock shaft to become inline with the ride spring so that we can add the forces and spring rates. If we mount the bumps on a fixture separate from the shock, mostly for convenience in making changes, then it would be a good idea to make sure that the motion ratio is the same for both the coil-over ride spring and the bump fixture.

Force Comes From The Sway Bar – The sway bar is a torsion spring, like we said previously. In chassis roll, it acts just like a ride spring. It also has a motion ratio for each side of the suspension. This motion ratio is

This road racing rear sway bar illustrates the idea that it is an added spring for the purpose of chassis roll. For these cars, with a solid axle rear suspension, the ride springs can be softer because the sway bar helps reduce chassis roll in the rear to match the front and rear roll angles for a balanced setup.

Summation – This is a very important Lesson. To review, the force needed to support the tire loads can, and does, come from different sources. These must be added together to come up with the force we will need to balance our setup.

The force can come from the Ride Spring, the Bump Device and the Sway bar as a spring. How we add those different springs together with their much different motion ratios will be explained soon. This is the art of modern day race car setup.

Exam - In The Context Of This Lesson:

The Concept Of Measuring Force Has Been Around For A Long Time?

1) True
2) False

Suspension Force Comes From Where?

1) The ride springs or torsion bars
2) The bump devices
3) The sway bar
4) All of the above

Motion Ratio As Discussed Is Defined As?

1) The wheel travel divided by the shock travel
2) The spring travel divided by the wheel travel
3) The sway bar arm travel divided by the wheel travel
4) The bump device travel divided by the wheel travel
5) 2, 3 and 4

The Force Needed To Support A Load Is Found How?

1) By multiplying the spring rate times the motion ratio
2) Dividing the spring rate by the motion ratio
3) By dividing the wheel load by the motion ratio
4) All of the above

A Sway Bar Can Be What?

1) An anti-roll device
2) Another ride spring
3) A torsion bar
4) All of the above

A Bump Device Can Be What?

1) Another ride spring
2) Linear in spring rate
3) Non-linear in spring rate
4) All of the above

Race Car Technology – Level Three
Lesson Fifteen – The Sway Bar As A Spring

In Lesson Thirteen, we discussed how the force needed to support the chassis at both ride height and at mid-turn is supplied in different ways and the sway bar is one of those ways. In the context of that Lesson, the sway bar contributed as an anti-roll mechanism and had a spring rate that caused the stiffening of the suspension that reduced the roll angle. Now we will look at a different use for the sway bar.

The effect we will be studying is only used or to be applied to non-symmetrical AA-arm front suspensions that run on circle tracks and only turn one way. I know of no good use for this with symmetrical chassis like a road racing car.

If you are interested in road racing technology, this won't necessarily apply, but it's not a bad idea to understand the principle. That is because in some types of road racing designs such as a circle track car that was converted to road racing, the linkages or designs may lend themselves to be A-symmetrical by accident. This Lesson may help you identify that condition and make the needed corrections.

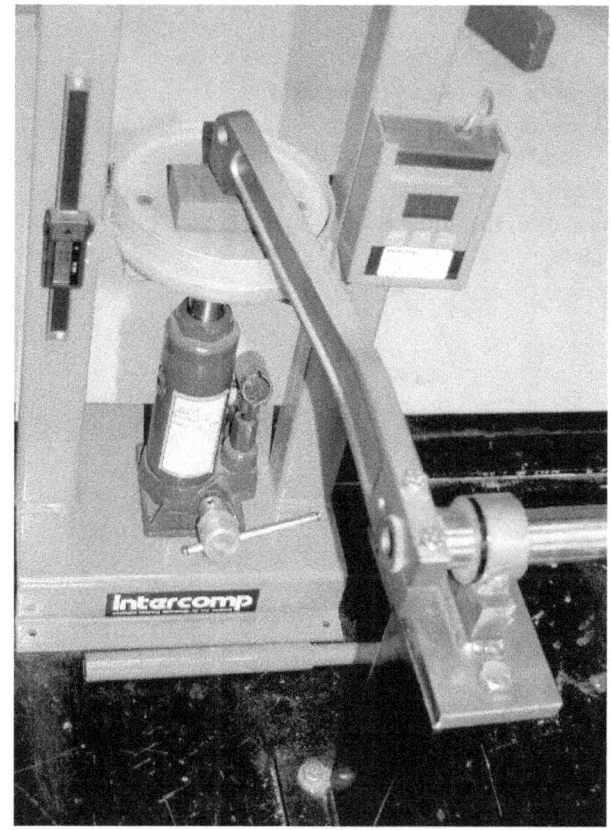

With the three-piece bar, we basically have a torsion bar and we can find the rate at the end of the bar using a fixture. We can also measure the force that bar produces on the car translated to the coil-over position with a Gale Force Sway bar tool.

We can rate a sway bar on a fixture, like this team did. That way we know exactly what the rate of the bar is at the end of the arm. This takes away all guess work. This one-piece bar will rate lower than an equal sized bar with equal effective length and equal length arms. That is because the hardness of the metal is less than with a three-piece torsion type of sway bar.

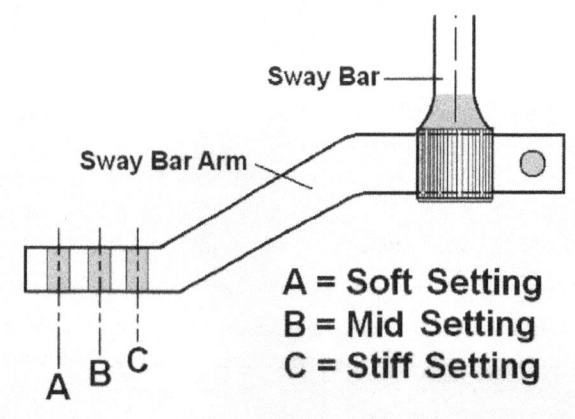

The rate of the bar is affected by the arm length. Some designs allow the team to change the length of the arm in the installation. A shorter arm provides more spring rate. If using this technique, be sure to move the control arm mounting holes too so that the link is parallel to the control arm and the sway bar arm.

Sway Bar In Roll – The sway bar, or anti-roll bar, was designed to reduce the roll of a AA-arm suspension by basically adding spring rate to each side of the suspension when the car rolled, and only when the car rolled. Well, that was the theory, and it works when the suspension is truly symmetrical.

So, if we have the same motion ratio on each side of the car, as the car dives, the sway bar has no anti-roll characteristics and the ends of the sway bar arms move the same amount vertically. There is no motion causing the bar to twist, so no anti-roll effect exists. That is the theory, but the application is a little different in most cases.

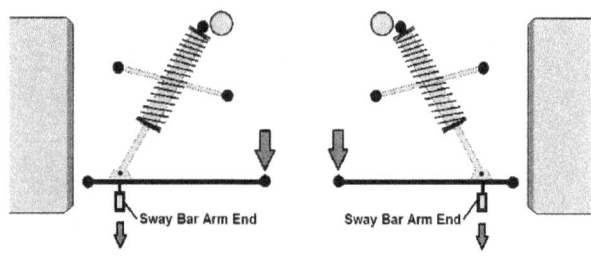

If the two sides are symmetrical, then the sway bar arms will move an equal distance as the chassis moves in dive. With this design, there will be no twisting of the sway bar caused by chassis dive, only chassis roll. This is common in road racing cars, but uncommon in circle track cars. If a circle track car has been converted to road racing, then the team must make sure that the motion ratios are the same side to side.

This is a concept little known or understood until very recently. With the advent of the soft spring setups where the car is made to run on bumps, and travel quite a bit, we discovered a phenomenon. We discovered that when we pushed the front end down onto the bumps, the sway bar, which was neutral at normal ride height, was now pre-loaded. The motion of moving the chassis down caused the sway bar to twist and become loaded some amount.

This is a major discovery and leads us to much more in the understanding of the sway bar and its contribution to the overall setup of the car and its contribution to the forces we need to support the cornering loads on the front tires.

The pre-load force of the sway bar can be adjusted using several different methods. This does not change the rate of the bar, only the pre-load and ultimate force of the system that includes the ride spring and the bump device.

For a race car using a one-piece sway bar, this is how it is adjusted for pre-load. The actual mounting point of the sway bar is moved up or down to twist the bar and add pre-load.

Sway Bar In An A-symmetrical System – For many circle track cars with offset chassis, and some that are called a perimeter, or symmetrical, the motion ratio of the sway bar mounts is not the same side to side. So, when the chassis moves vertically, and separate from the roll motion, the two ends of the bar do not move the same distance. In most cases, the left arm end moves less than the right arm end.

Relative to the chassis, the right arm end moves up more than the left arm end and this simulates what happens with chassis roll. Therefore, the sway bar supplies an anti-roll force with just chassis dive. Then when the chassis rolls, the bar is twisted to a greater extent.

Both of the actions, twist from chassis dive and twist from chassis roll, provide a force to the right front suspension that is added to the ride spring force and the bump force. If we only consider the ride spring and bump forces, we will be leaving out what could be a significant contributing force that helps support the mid-turn loads.

We have talked about recording the shock/spring travels so that we can simulate the force generated by the ride spring and the bump device. But what about the sway bar force, it needs to be measured too or we won't be able to properly adjust the ride spring and bump forces to what they need to be.

Suppose we need 2300 pounds of force to support 1500 pounds of loading on the right front tire. If the sway bar is supplying 400 pounds of force all by itself, then if we set the ride spring and bump combo to supply 2300 pounds of force, in reality we will be creating a force that is 400 pounds too high and our setup will not be balanced.

So, we need to know not only the force generated by the sway bar in chassis roll, but also in chassis dive. There are measuring devices that can provide the actual force the sway bar is generating when we put the chassis at the mid-turn attitude that simulates both chassis dive and roll. Then we can subtract that force from the total force needed to support the tire loading at mid-turn to then know how much force we will need from the ride spring/bump combination.

The Sway Bar Use As A Spring – So, far, we have talked about how the sway bar might act as both a spring in chassis dive, and as a spring resisting chassis roll. Let's look deeper into how that is happening and see if we can utilize this property to our advantage.

The motion ratios can be different in an offset chassis because the lower control arms are different lengths. If the end of the sway bar arm is mounted the same distance from the ball joint for the left and right lower control arms, then due to the difference in lengths, the ends will travel different amounts in chassis dive.

If the sway bar arm ends are mounted at different distances from the ball joints for the left and right lower control arms, then this difference in travel of the arm ends becomes greater. We can actually increase the dive force amount that the sway bar contributes by moving the mounting points of the sway bar arm ends.

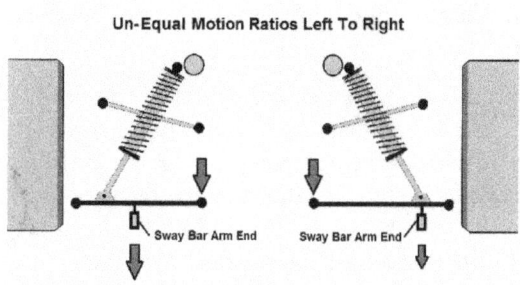

When we move the sway bar arm end towards the chassis mount, we change the motion ratio for that side. In chassis dive, the left arm end (left side of the sketch) will move more than the right side. This causes the sway bar to twist and puts force onto the right lower control arm. This force is added to the force of the ride spring and the bump at maximum mid-turn travel of the suspension. The total force needed to support the RF tire load at mid-turn is a product of sway bar force, ride spring force and bump force.

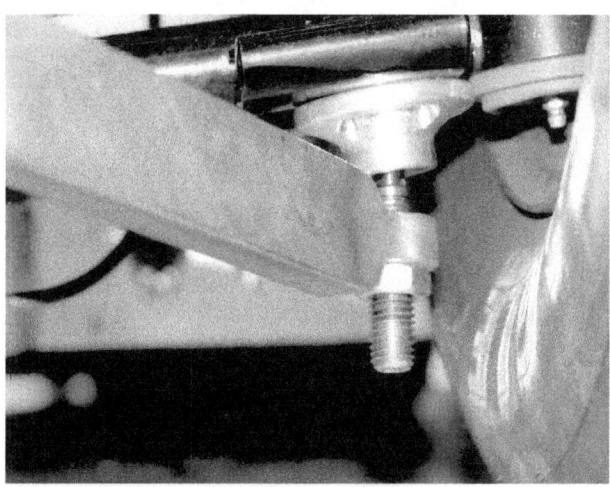

To change the motion ratio in this left side lower control arm, we need to install an arm with less top view angle that will move the contact point in towards the chassis mount. This decreases the motion ratio causing the left arm end to move a greater distance than the right arm end when the chassis dives.

If we move the left side arm end in towards the inner chassis mount, then the motion ratio becomes less. Let's exaggerate that so we can better understand this principle. As the chassis dives, the sway bar itself moves down with the chassis. The arm ends move much less in a normal system. So, let's take that a little farther.

Suppose we mount the right sway bar arm end right under the ball joint. Now this arm will not move vertically as the chassis dives because the ball joint doesn't move either. Now if we mount the left sway bar arm end under the left lower control arm chassis mount, it will move the same distance as the chassis and sway bar.

Now, with this exaggeration, the right arm end does not move and the left arm end moves with the chassis. Suppose the chassis were moving 3.00 inches in dive? Then we would be twisting the sway bar by three inches. If the sway bar rate were say 250 ppi, then the force generated would be 750 pounds. That is a significant amount of force. If the bar were a larger sway bar with a rate of 766 ppi, then the sway bar alone would generate the 2,300 pounds of force needed to support the tire loading at mid-turn.

This example is extreme for sure, and we would never mount the left side sway bar arm end in this manner. But what if moving the arm end in towards the chassis mount by an inch or two supplied the added force we need for some types of setups? Then we might just be on to something.

The Left Bump Only Setups – In a bump setup where the two sides of the front suspension had bumps installed, there would be plenty of force available to support the right and left front tire loads at mid-turn. Some race teams only run one bump and use it on the left side of the suspension.

Then the single bump team might run a stiffer RF spring than the LF spring, but never as much spring rate as the left side has when we combine the ride spring rate and the bump spring rate. So, what holds up the car? The sway bar does.

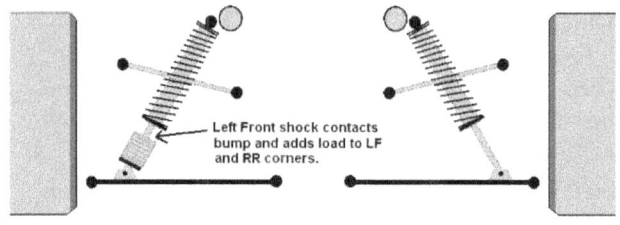

Asphalt Single Bump On Entry

Left Front shock contacts bump and adds load to LF and RR corners.

When a race car is designed to use only a left side bump (left side of sketch), there will be plenty of mechanism to provide force for mid-turn support. But at the right side, with no bump and a soft ride spring, there is nothing to provide the force needed to support that corner of the car at mid-turn. We need to install a larger sway bar and an un-equal motion ratio design so that as the chassis travels, the sway bar will provide a force along with the ride spring that equals what is needed to support the tire loads at mid-turn. This is why single bump setups require a large sway bar, and double bump setups require a smaller sway bar.

From what we have learned, an A-symmetrical front suspension that has different motion ratios for the sway bar arm ends will produce a dive spring rate which can assist the RF ride spring in providing sufficient force to support the RF tire loading.

For our sample single bump car, we have a 150 ppi ride spring in the RF, a tire static loading of 650 pounds and a dynamic tire loading of 1500 pounds. For the ride spring to support the 650 pounds of static tire loading, it must travel 6.5 inches or so. We divided the 650 by our motion ratio squared of 0.660 to get that travel amount. This represents the pre-load on the ride spring at ride height.

The additional travel amount that spring would need for a 1500 pound mid-turn loading would be 8.5 inches more, and we just don't have that much distance available. Three inches of chassis travel is only 3.7 inches of shock travel. We'll bottom out long before that spring supplies enough force to support the car. So, we need more force.

If the sway bar twists a half inch at the arm ends in three inches of travel, then what size sway bar do we need to supply the added force to support the RF tire loading? The three inches of travel will give us 3 x 150 = 450 pounds of added force from the ride spring. We need a total of 2,300 pounds of overall force to support the 1500 pound so tire loading.

If the sway bar provided 985 pounds at ride height plus another 450 pounds from 3 inches of travel, then it is supplying a total of 1,435 pounds of force. We need a total of 2,300 pounds of force, or another 865 pounds.

A 1 15/16" sway bar is rated at around 1,730 ppi (pounds per inch). If the sway bar arm ends moved a half inch, then that bar would provide half the full inch rate, or 865 pounds of force in three inches of travel and exactly what we needed in additional force. Now we have the total force we need to support that tire loading at mid-turn.

Re-Cap - To recap all of this for you, the ride spring provided a total of 985 pounds (at ride height to support the tire load) plus 3 x 150 = 450 pounds (from chassis travel) for a sum of 1,435 pounds. We needed a total of 2,300 pounds of spring force to support the mid-turn tire loading, so we need 2,300 – 1,435 = 865 more pounds of force than we are getting from the ride spring alone.

Since a 1 15/16 inch diameter sway bar produces 1,730 pounds in one inch of arm movement, and the chassis dive produced a half inch of arm movement in 3 inches of dive, we divide the ppi rate of the bar by two to get 865 pounds of force, or exactly what we need added to the ride spring force to support the RF tire at mid-turn.

If our sway bar arm mount on the LF lower control arm was farther in from the ball joint towards the chassis mount, then the bar would twist more and we would then require a smaller sway bar to generate the same 865 pounds of force.

Suppose we had one inch of sway bar arm travel from the 3.0 inches of chassis dive, then we could use a sway bar rated at 865 ppi, or very close to a 1.50 inch diameter sway bar. Now we are beginning to understand why some teams run larger or smaller sway bars than others. It's all in the design and mounting points we have on our cars.

to 1.250" sizes. For single bumps setups, much more force, or spring rate, is needed and we are talking about sizes above 1.500" up to nearly 2.00" diameters. The more un-equal the motion ratios, the less sway bar rate we will need for single bump setups.

Why We Use The Sway Bar As A Spring – This is exactly why we use a large sway bar for single bump setups. For a double bump setup, we already have plenty of force that is provided by the RF bump device, so we must use a much smaller diameter sway bar that has much less rate.

By the motion ratio rule, we can change the left side motion ratio to adjust the sway bar spring force at the RF corner. Professional race teams now experiment with different LF motion ratios to tune the forces at the RF corner rather than change the sway bar size.

Summary – Much of what we know comes from the application of new technology. One thing leads to another and at some point, different approaches and different results add up to a better understanding of the total package.

What you have just learned is less than a year old. It helps us understand how to setup our race cars. Now we know why we need a larger sway bar for the single bump setups, and why we must use a much smaller sway bar for double bump setups.

Sample Sway Bar Spring Rates

Sway Bar Rates: (Sway bar Effective Length = 36.0" and Arm Length = 11.50")

0.750" w/ solid bar	= 72.0 ppi
0.875" w/ solid bar	= 133.0 ppi
1.000" w/ 0.50" hole	= 212.0 ppi
1.125" w/ 0.75" hole	= 264.0 ppi
1.250" w/ 0.75" hole	= 481.0 ppi
1.375" w/ 1.0" hole	= 584.0 ppi
1.375" w/ 0.875" hole	= 677.0 ppi
1.500" w/ 1.125" hole	= 812.0 ppi
1.500" w/ 1.00" hole	= 922.0 ppi
1.625" w/ 1.25" hole	= 1027.0 ppi
1.625" w/ 1.125" hole	= 1244.0 ppi
1.750" w/ 1.250" hole	= 1573.0 ppi
2.000" w/ 1.625" hole	= 2046.0 ppi
2.000" w/ 1.500" hole	= 2478.0 ppi

This chart represents the spring rate of three piece sway bars, or torsion bars, for the dimensions shown. We can see how the rates increase as the sway bar diameter sizes increase. Double bumps setups can utilize much smaller diameter sway bars in the 0.750"

Exam - In The Context Of This Lesson:

Non-Symmetrical AA-arm Suspensions Are Common With?

1) Formula race cars

2) Circle track offset chassis

3) Circle track perimeter cars

4) Road racing cars

We Find The Working Rate Of The Bar By?

1) Measuring the diameter of the bar

2) Measuring the arm length

3) Rating it at the end of the arm

4) Finding the length of the sway bar

A Sway Bar Can Produce A Force In What?

1) Dive only

2) Roll only

3) Dive and roll

4) All of the above

A Symmetric AA-arm Suspensions Produces Force When?

1) The car rolls

2) The chassis dives

Non-Symmetrical AA-arm Suspensions Produces Force When?

1) The car rolls

2) The chassis dives

3) 1 and 2

For Single Bump Setups, We Need What?

1) A Non-symmetrical suspension

2) A large sway bar with a high rate

3) Additional Spring force at the right front

4) All of the above

For Double Bump Setups, We Need What?

1) A more symmetrical suspension

2) A smaller sway bar with a lower rate

3) Less sway bar force at the right front

4) All of the above

Race Car Technology – Level Three
Lesson Sixteen – Aero Downforce & Tire Loading

The goal of this school is to give you, the racer or want-to-be racer, real world knowledge about the subject matter that matters. We offer views on older technology just so you can get a perspective about whatever it is we are discussing. As far as aero technology is concerned, we feel that a complete picture of aero as it applies to race cars cannot be presented without addressing a few common, but erroneous assumptions.

In years past, some tracks allowed innovation in the body design and shapes. This late model that ran at New Smyrna Speedway in the mid-1970's utilized a high downforce nose using Flat Plate aero, and a low drag, low mounted roof. The tail was appropriately called a "whale tail" due to its shape and how large it was. We see utilization of flat plate aero as well as low pressure under the hood in this design.

I have formed my opinions by studying aerodynamics and talking to persons who live aerodynamics. And I have come to agree with them for some parts of aero and have offered my take on other parts of aero technology.

When you design an airplane or a race car that you will ultimately have to fly or drive, you must make sure you have a real world understanding of the subject of aerodynamic engineering. The classroom or cozy control room in a wind tunnel presents little risk and there is nothing to prove the results wrong, until you "fly" the airplane or race car.

Wind tunnels will give you tendencies and probabilities only, and that is why major airplane design companies have test pilots. Not one airplane ever designed and "flown" in a wind tunnel ever flew in test flights the exact same way. It's the same with race cars. Otherwise, every multi-million dollar F1 team would have the perfect aero properties in their cars design, and we know for a fact they don't.

So, the data we collect when using a wind tunnel has to be translated somewhat into information that will more closely resemble what happens when we "fly" the airplane or race car. There is various editions of software that will get that done and there is computational fluid dynamics software that will attempt to simulate what the wind tunnels, and real world flying, does.

These computer programs are not perfect and there is still room for practical knowledge of aerodynamics for the enthusiast and applications engineers in the discipline of aerodynamics.

Wind Tunnels Have Errors – The whole premise of a wind tunnel is to move air at a given speed into and over an object to measure lift, down-force, drag, etc. In a wind tunnel, the air is accelerated to some predetermined speed and the stationary object, the race car, is bombarded by the oncoming, energized air that has a lot of momentum. This is not what happens in nature with airplanes and race cars. They do the opposite by traveling at speed through relatively still air that has no energy or momentum. This is a major point to make. This is how you develop a practical knowledge of aerodynamics.

Some aero engineers will tell you, "Well, it's the same thing." No, it is not. Air moving at 60-100mph has a lot of energy and does not want to deviate from its path, kind of like a baseball traveling the same speed. When this energized air hits an object, the way it moves around that object is much different than if that object were moving through still air at the same speed. It has to.

Most veteran aero engineers, if they are honest, will tell you that the primary difficulty with wind tunnel data is converting it to reality. There exists some very expensive Computational Fluid Dynamics, or CFD for short, software programs that attempt to convert the wind tunnel data and make it portray what happens in the real world. The conversion is very complicated and obviously not perfectly accurate. And again, that's why we have race tracks and test pilots.

The other thing that causes errors in the data we get from a wind tunnel test is the close proximity of the walls. Most wind tunnels are too small because it is cost prohibitive to build it big enough so that the walls don't influence the results. Air is disturbed well beyond the distance to the walls in most wind tunnels when using full scale models. The presence of the walls restricts the movement of the surrounding air and causes a compression effect which alters the results.

The only "wind" tunnel that comes close to reality is the one Chip Ganassi is said to own called Laurel Hill in Pennsylvania. It is actually a tunnel through a mountain. There a test vehicle is run for a mile through the tunnel, through still air, and the pressure distribution that produces drag and down-force is measured. At least the test object is traveling through relatively still air just like an airplane or race car really does. While the tunnel still has walls that are too close, the results are much closer to reality.

The ideal aero testing for race cars would be done on a smooth flat surface outdoors at speeds that replicate the speeds the car would experience on a race track. Then the load and pressure sensors would record the aero influence on the car, much like test flights do with airplanes.

That Being Said – Now that we have that out of the way, let's get into how aero really works. The age-old depiction of an airplane wing, and one that I have used in the past, is not perfectly correct. Yes, the air traveling over the airplane wing travels farther and faster and thus may have less pressure than the air traveling under the wing.

This is often called the Bernoulli principle and it supposedly provides lift. But it really doesn't. The Bernoulli theory used a closed container and moving air. The air a race car flies through is not moving. It is being pushed out of the way. So, to say it flows is not exactly correct. It moves laterally and vertically as the race car speeds through it.

The Problem With Bernoulli – The Bernoulli principle states that, "As the velocity of fluid flow increases, the pressure exerted by that fluid decreases". This may have some validity in a wind tunnel where the "fluid" air is moving, but in the real world, our race cars and airplanes move through air that is not moving at all relative to the motion of the airplane.

What the air in the real world does is lay around until an airplane or race car moves through it at a high speed. The air is then pushed out of the way and displaced whereby it pushes against the surrounding air that is itself relatively still and not moving. Some of the displaced air is pulled along with the vehicle moving through it until it is released and allowed to flow back in a direction where it was before it was disturbed.

There is no way a wind tunnel can re-create the action we just described. It is so different from the scenario of moving air flowing over a still object that the two are completely different in the dynamics of the event.

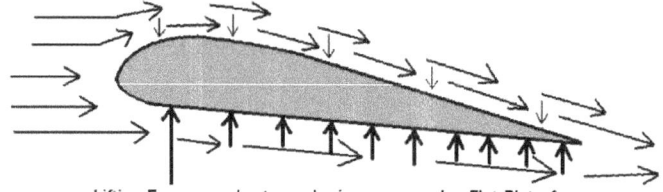

An airplane flies due to lift, but one seldom talked about component of lift is Flat Plate aero affect. It is the lift provided by the flatter underside of the wing moving through air at a slight angle that counts for a large percentage of the lifting component. The low pressure on the upper side, which is often erroneously explained using the Bernoulli's principle, reacts with the higher atmospheric pressure on the underside to provide the other part of lift. We can utilize the FP aero affect alone on our race cars where the conditions for Bernoulli's principle does not exist.

Most every airplane cruises with a wing attitude where the underside of the wing is at a small angle (the front higher than the back) to the air it is moving through. So, it acts much like a water ski to assist in holding up the airplane. This is called Flat Plate (FP) aero.

This early-2000's design of prototype Grand Am race car utilizes flat plate aero on the front of the car. These cars were very successful during their rein. They produced a lot of downforce.

Many aerodynamicists tend to discount this affect, but it can be a large percent of the total lift component. This is where we get into an important area of race car aero design.

In a race car, we may have flat plate areas where we can develop down-force to increase the grip of the tires. That is called FP aero down-force, but what about other areas of the car that are flat? What about the sides? We'll get into that in a minute.

In race car aero design, one often overlooked component of aero down-force is what is called Flat Plate Aero. On this outlaw late model, the front nose is angled down at about a 45 degree angle to create Flat Plate down-force. We'll explain how that works.

Down-force From Flat Plate – The sloped front of an Outlaw late model, the nose and hood on a '70's era super late model, the nose on a modern dirt late model and the wing on a sprint car all offer FP aero down-force possibilities.

The modern dirt late model race car utilizes flat plate aero not only at the front, but along the outer sides. These cars use rear steer in large amounts to get the rear of the car out and into the air. This provides flat plate aero sideforce that pushes the car towards the inside of the track, countering the lateral forces trying to push the car off the track.

This nose on an asphalt modified is relatively small in area, but effective. The raised parts on the outside edges help keep the air on the flat plate and helps prevent it from spilling off the sides.

One of the early proving grounds for short track racing was the annual February Speedweeks races at New Smyrna Speedway in Florida. Back in the late 1970's and early '80's, teams from all over the country would show up with "stock" bodied super late model cars that adhered to the strict rules of their perspective sanctioning bodies only to discover that there were no body rules at NSS. The local teams already had their car bodies tricked up.

Those who remember noticed that after about the second or third night of racing in the nine night series, most of the cars had been transformed into sleek, wedge shaped oddities with huge fantail rear spoilers attached. This is what we now understand to be an effective use of FP down-force.

With each of the examples, we may, or may not, have a low pressure on the other side of the FP, but the FP still generates its own force. The angle is critical and relative to the speed of the vehicle, just as it has proven to be with airplane design, but up to a certain angle of about 15-20 degrees, a lot of force can be developed.

As for the wing on a sprint car, most of the down-force from a sprint car wing is FP derived from air flowing over, and being deflected up, off the top flat plate. It acts much more like an outlaw nose. The shape of the bottom of a sprint car wing being curve like the top of an airplane wing, is there to transition the air flowing around the wing so that there is less drag.

And you can have too much angle in that wing. As the speed increases, the wing angle must be reduced in order to have maximum down-force while not creating too much drag. This is not unlike with an airplane where we trim the attitude of the plane and wings, and retract the flaps, as the speed increases.

A typical winged sprint car uses flat plate aero on its wings both as downforce as well as for side force from the side plates running at an angle to the direction of travel of the car. These wings can produce quite a bit of downforce helping these cars maintain very high speeds through the turns.

Flat Side Lateral Force – The concept of FP aero can be applied laterally in addition to vertically to assist in helping the car to turn. If a race car has relatively flat body work on the side of the car towards the outside of the turns, then if the car is run at an angle to the direction of travel, it can produce a force in a direction opposite to the centrifugal, or lateral forces trying to push the car off the track.

Cars where this can be a benefit are easy to pick out. Late model dirt cars, dirt modifieds, Outlaw late models are just a few. And, if we look at the large sideplates on a sprint car wing, we see where FP aero can be used if the car runs at just the right angle through the air.

In the case of the dirt late model, we know these teams rear-steer the car so that the rear tires are running outside the front tires. This puts the large flat side of the car at just the right angle to the air it is moving through to produce FP forces that are opposite of the lateral forces. Not equal to, but opposite in some amount. The tires still have to do the rest of the work, but the car goes faster with this FP advantage.

It's a similar situation with sprint cars. I have watched Outlaw sprint cars qualify at Volusia during Speedweeks at upwards of 135 MPH average. The weight of the car on the tires cannot hold the car in the turns at that speed. There must be FP aero down-force and side-force helping to keep the car on the track.

Spoiler Down-force Vs. Drag – When racers run tests with their cars in a wind tunnel, they experiment with spoiler angles. Sometimes the results can be miss-interpreted. Here is what I mean. A more vertical spoiler will produce much more drag than one angled at say, 55 degrees off of vertical.

The wind tunnel operator must compare changes in loading on each axle with the total loading of the car. So, let's say a vertical spoiler added 50 pounds to the rear axle, but removed 45 pounds from the front axle. The result is only 5 pounds of added down-force combined with a lot more drag, and represents more of a displacement of loads due to the leverage of drag. If the total loading on the car does not increase, you have no gain. You might as well take that 5 pounds of weight from the front of the car to the rear and remove the spoiler.

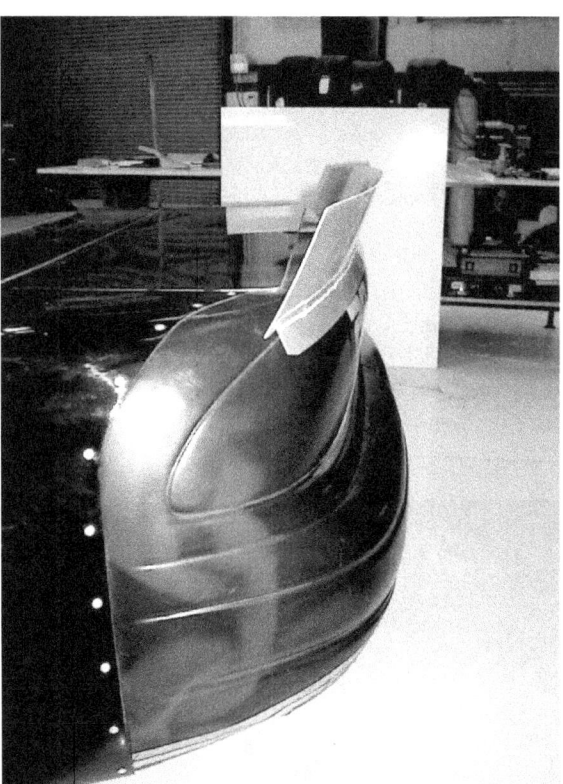

There has been much discussion about spoiler angles and their effectiveness. High angles (from horizontal) result in a lot of drag and very little actual downforce. This drag pulls on the rear of the car and does displace load from the front to the rear. Any increase in rear axle loading can be mostly attributed to load coming off the front axle and transferred onto the rear axle.

With the 55 degree spoiler angle, we might see a 40 pound increase in the rear loading and a 15 pound decrease in the front loading due to the same leverage affect from drag, but in this case the drag is much less. In this example, we see a total increase in the vehicle loading of 25 pounds from actual aero FP down-force.

Traditional Low Pressure Downforce – In a more traditional sense of looking at aero down-force, there is the phenomenon of high pressure opposite low pressure. The shape of the body on our race car can help produce low pressure inside the body panels to create even more down-force. What we need to do is design our bodies so that we can direct some of the air we are driving through around the car in order to vacuum air out from within the areas inside the body.

If we can use the movement of air that is being pushed out beyond the sides of the car to help vacuum air out of the engine compartment under the hood, we can lower the pressure along the underside of the hood. A higher pressure on one side of an object will push that object in the direction of the lower pressure, or high towards low.

To accomplish this, teams use wider, angled front noses that will direct the displaced air around the sides of the car beyond the wheel wells in such a way that a low pressure area is created just outside the wheels. Air then moves out of the engine compartment to fill this "void" and the pressure under the hood is reduced.

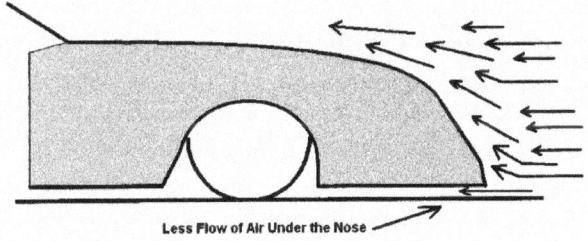

It is common in today's racing to position the race car low to the race track in the front. This accomplishes several beneficial things. First off, the Center of Gravity is lower, there is very little camber change as we previously discussed, and a low nose section cuts air off that would otherwise invade the under side of the hood where we are trying to create a low pressure area. If we can keep air from getting under the car, we can realize a greater low pressure effect under the hood.

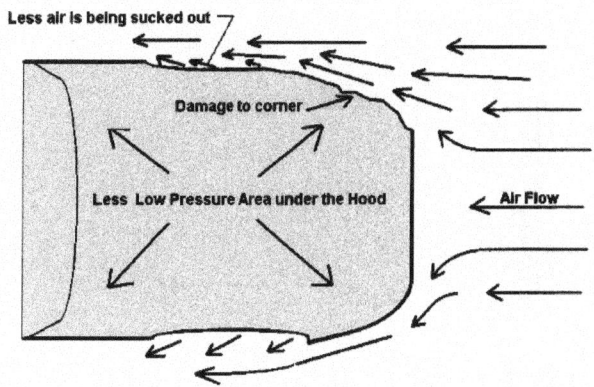

This illustrates how air being routed around the nose of a race car can cause air under the hood area to be pulled out into the low pressure area around the wheel wells. Any damage to the nose and fenders defeats this purpose.

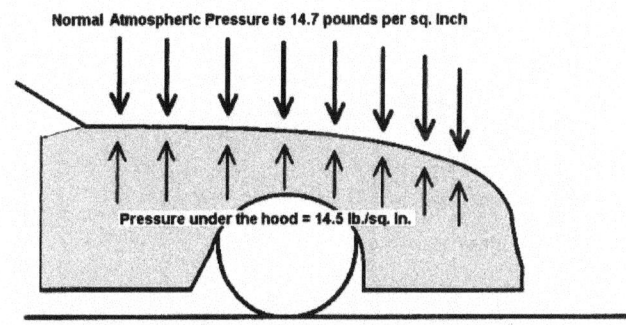

A simple calculation shows that for a small decrease in atmospheric pressure under the hood, we can see upwards of 300 pounds of downforce. This is distributed equally onto the front tires at about 150 pounds each. That is a lot of load gain and added traction.

The average atmospheric pressure at sea level is 14.7 pounds per square inch on all sides of an object, even on our bodies. If we reduce the pressure under the hood to 14.5 PSI, a drop of only 0.2 PSI, over an area of just a square yard we would generate about 260 pounds of down-force ($0.20 \times 36^2 = 259.2$ pounds).

Rear Aero - At the rear of the car, we can manipulate the shape of the spoiler, the rear window posts and the body just in front of the rear wheel wells. By routing the air that is flowing past the sides of the car out and to the sides of the rear wheel wells, a similar suction effect takes place to create a low pressure area under the rear deck. There are obvious limits as to body shape in this area, but a little reshaping can help.

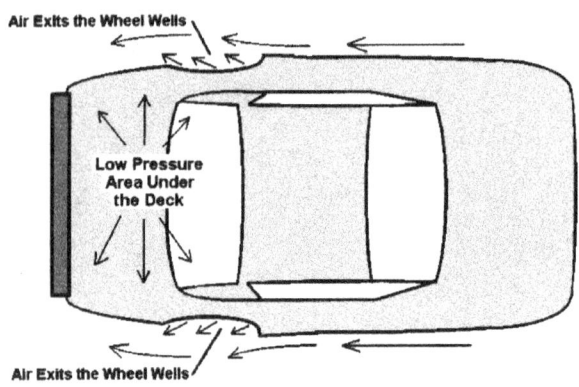

At the rear of certain bodied cars, we can create low pressure down-force by routing air out of the rear cavity under the rear deck. This gets tricky with current strict body rules, but most tech officials are looking elsewhere for "cheating".

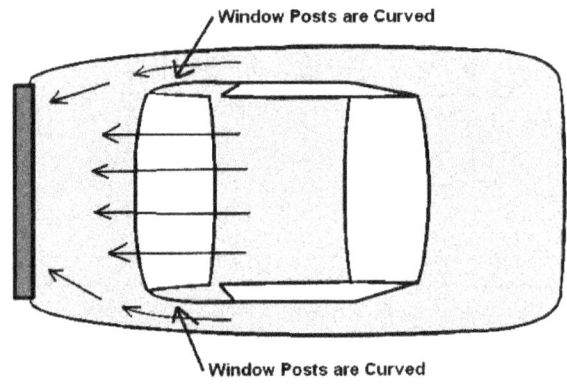

Another trick used by some professional race teams is to shape the window post areas so that they are curved allowing air to flow around the post and onto the rear spoiler. The greater amount of flowing air going onto the spoiler instead of over and around it causes more rear downforce.

Race tracks that are longer and have more banking require less down-force and would benefit from reduced drag. We can rethink how the air flows past the roof (green house area) and onto the rear spoiler. If we reshape the post that connects the rear window with the side window openings, we can direct air away from the spoiler and greatly reduce aero drag.

The modern day race car has a low attitude combined with a nose shape that provides low pressure under the hood, flat plate aero properties across the top of the hood, and rear air flow that enhances rear deck low pressure.

Summary - The key goals with stock car aero design and development is to create a body shape and running attitude that will provide more down-force and side-force to enhance turn speeds, produce less drag, and promote a more balanced race car. We can now see where and how we can utilize Flat Plate aero to increase down, as well as side, force.

Everything we do has to be done within certain limits. If we work hard to develop 600 pounds of down-force on the front end and the rear is not able to keep up with that high amount of grip, then the car will be loose and nobody can drive a loose car fast. So, there are limits to how far we can go.

Work towards a good balance of front to rear aero down-force to help produce more overall grip. Do not overdo your efforts to help the car aerodynamically at the expense of handling efficiency. Make sure the basic chassis setup is balanced, and then the combination of both aero down-force and handling will enhance the on-track performance.

Exam - In The Context Of This Lesson:

Modern Aerodynamics Is The Study Of?

1) Air flow across a wing or race car.

2) How air reacts when an object is moved through it

3) The Bernoulli principle of low pressure vessels

4) Wind tunnel data

How Is Moving Air Different Than Still Air?

1) Moving air resembles how a race car reacts

2) Moving air has less energy

3) Still air has no energy unlike moving air

4) It is how a wind tunnel works

Which Of The Following Are Deficiencies Of A Wind Tunnel?

1) The air is moving and has energy

2) The walls are too close to a full scale car

3) The data cannot be accurately converted to rear world numbers

4) All of the above

Flat Plate Aero Is Created When?

1) The air is moving and flows over a flat plate

2) The flat plate panel is moved through still air and is pushed in some direction

3) Low pressure air is on one side of a body panel

4) We have flat sides on the race car

Downforce From Low Pressure Is Caused By?

1) The air being evacuated out of a closed area

2) A higher pressure being created opposite a low pressure area

3) Air flowing around an opening outside a closed in area

4) All of the above

A Low Nose Helps Performance and Efficiency By?

1) Keeping air from invading the area under the hood

2) Providing a higher angle of the nose flat plate to produce more downforce

3) Creating a lower Center of Gravity

4) All of the above

Race Car Technology – Level Three
Lesson Seventeen – Practical Applications – Part One

All that we have done over the course of Levels One through Three has led up to this Lesson. This might be considered the holy grail of race car technology by some. It's not the final chapter, but one of the most important because here we will learn how to develop the actual setup for the race car.

Our approach to setup is designed to be the most efficient and quickest way to get fast and end up with a truly balanced setup, something we stressed all along. This approach may differ from some others in that we will tune for the Mid-turn portion of the track first, then tune the entry and then the exit. We do this for a reason.

The balance we seek in order to have the perfect setup will make our mid-turn performance the best it can be. Once we find that, our entry and exit balance will be much better also. Each one of these cars has made changes to find the mid-turn balance and each one probably got there a different way. There are many different ways to arrive at the ideal setup balance and we'll tell you what those are and how to make them.

The balance concept for dirt cars is arrived at and put together a little differently. Dirt cars need to be made to work on very slick surfaces at times. So, the setups vary as to the balance, whereas, on asphalt, the grip level of the racing surface remains fairly consistent. Here we see the 15 car has a rear roll angle that is much higher than the front roll angle. This condition will cause the left front wheel and tire to be lifted off the race track. This is a great illustration of the rear out-rolling the front.

The setup adjustments for a formula or prototype race car that has AA-arm front and rear suspensions is more limited than with a circle track car that only needs to turn one direction. The formula car must be symmetric in spring rates left to right. These car work mostly with the sway bar mechanism and front to rear aero balance to tune the setup balance.

The "tools" we will describe using are the tools we use in real life. And, these are the tools many race teams use successfully to setup their race cars, in both amateur and professional series. This is a disclaimer. Some of the tools we describe and use are tools we produce and sell. Some of these tools can be obtained from other sources, some of them cannot be sourced anywhere else. So, we have to use these tools or we would be doing you a dis-service.

To clarify, one of the tools is a force measuring rig, or machine. We use the Gale Force Suspension rig because Bubba Gale is my partner in this school and that rig works very well. There are other rigs available that do basically the same thing, but when you understand how the GF rig works, you will understand the principle behind it and that is what counts.

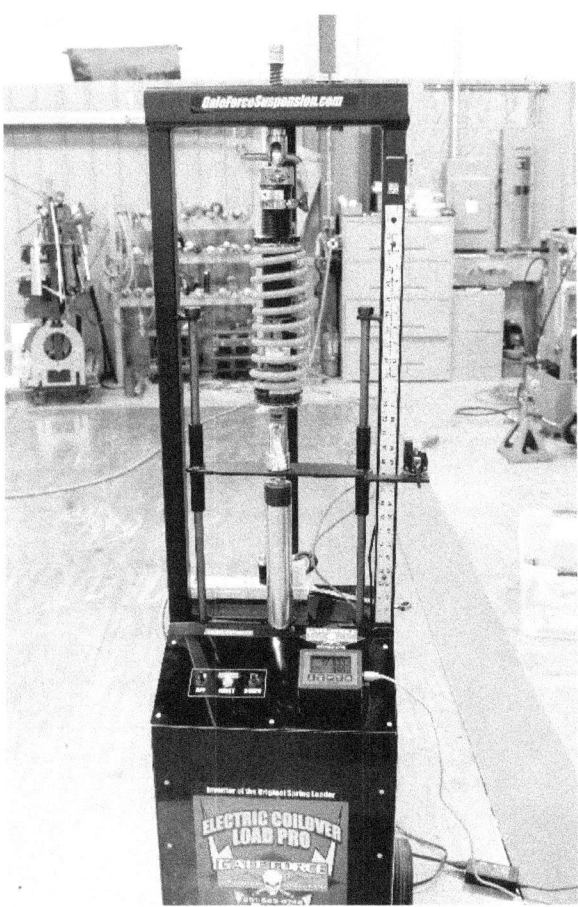

This is the Gale Force Suspension Load Pro spring loader. It measures the force produced by the spring and bump combination and uses the spring travel that is recorded on the race track. We will be using this tool to explain setup concepts because this type of tool is widely used in todays racing environment.

The other tool is the Chassis R&D Asphalt Setup software program that I developed in 1996. As far as I know, there is no other tool like this on the market. This tool has been used by me and thousands of racers over the past twenty years. If there were other tools like this that did what this tool did, I would tell you. There isn't, so we'll use this tool.

The setup tools help you to understand exactly what the goals are for ideal setup, how to get there, and really helps us to understand the significance of each setup change. As we go along with this Lesson, you'll understand more about that and come to appreciate that we use these tools in this school. This is where the racing industry is right now.

The Chassis R&D Setup software program was first offered commercially in 1996. Since that time, many thousands of programs have been used by racers all over the world. It calculates a predicted roll angle for the front and rear of a race car. On asphalt, we match the two roll angles to arrive at the perfect setup balance. We use this tool in the RCT courses because it is the only one of its kind available.

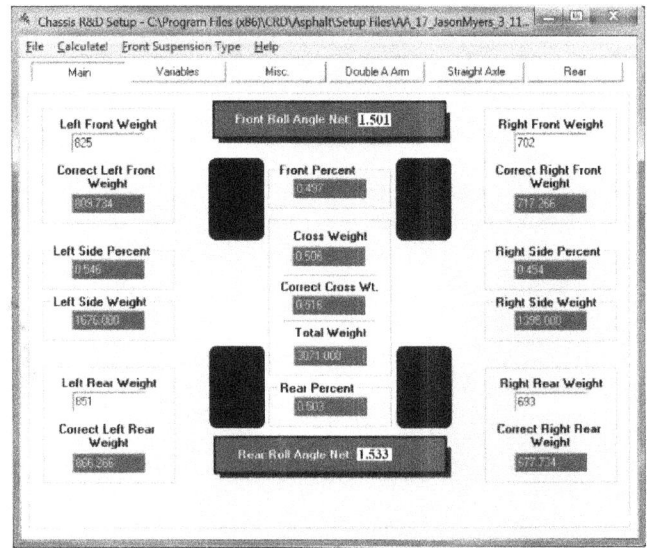

The CRD Asphalt setup software is very simple in concept, but very effective in the information it provides. The two answers it gives us are the Front and Rear Roll Angles. When we match these two angles, the two suspension systems are desiring to roll to the exact same angle and are working together. This is the balanced setup we are trying to achieve.

Where Do We Start Out? - Now that we got that out of the way, let's get started. The reason we tune the middle of the turns first is that when we achieve a balanced setup for the mid-turn, it often serves to balance the entry and exit. And, we always tune the spring and sway bar setups before we ever start tuning with the shocks. That's not to say that spring changes won't help entry and exit, it's just that those kinds of changes will be made while also keeping the mid-turn balance from changing.

The shock Lesson comes in the next Lesson and any shock expert worth their salt will tell you to first work out the basic setup before beginning to tune with the shocks. And that is what we are going to do. To explain, at mid-turn, the shocks and springs are in what is called a "steady state" condition. That means that for a short period of time, they are not moving. The car is not slowing and it is not accelerating and it is not moving vertically. That is the definition of steady state.

The middle part of the turn is also the slowest part and where the lateral G-forces, or centrifugal forces, are the greatest. It is here where the two suspension systems are rolled and compressed to the greatest extent. All of the way towards the mid-turn point, the G-forces are building and the radius is getting shorter.

The way we calculate the G-force is by using speed and turn radius, the more speed and the tighter the radius, the greater the G-force. Most overall turn layouts are elliptical in shape. That is, from the time the car first begins to turn until it gets to mid-turn, the radius is constantly decreasing. Then as the car begins to move away from the mid-turn, the radius constantly increases. If we plot this line, we will see that it is an ellipse. This is a very important concept to understand, especially for the drivers.

We will do some exercises in this Lesson where we apply gradually increasing G-force and see if our balanced setup stays balanced in the parts of the turn that is not mid-turn. The fastest driver will maintain a speed on entry that matches the turn radius providing the most performance possible. More on that later.

Mid-Turn Balance – The definition of mid-turn balance means that both ends of the car are trying to do the same thing. The "same thing" is defined here as roll to the same angle. If we take all of the effects placed on each suspension, like lateral force, gravitational forces, banking angle down force, etc. then each suspension system, front and rear, will want to roll to a specific angle.

Each suspension system is not free to roll to the exact angle it desires because in most cases, a stiff chassis connects the two suspension systems. If we read the chassis roll angle by doing a calculation using shock travels, it will not tell us what the desired roll angles are for each suspension, only the average of the two desires. Here, I will say that again.

If we read the chassis roll angle by doing a calculation using shock travels, it will not tell us what the desired roll angles are for each suspension, only the average of the two desires.

The setup software we told you about starting out in this Lesson will calculate the desired roll angle for each suspension system. This is the angle a suspension system would produce if that suspension system were not attached to the other suspension system. When we calculate both roll angles, we can then match the desires of the two systems by making changes to affect the roll angles, independently. We don't know of any other way to do this.

In the following examples and situations, we will be calculating the roll angles of the front and rear suspensions to demonstrate the magnitude of the effects of spring and sway bar changes, which is very valuable in working out our setups.

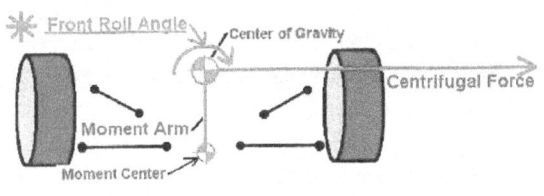
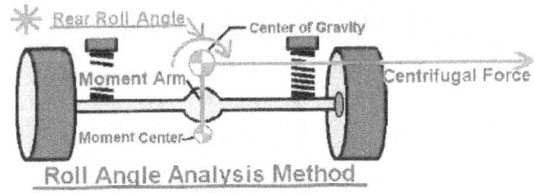

Roll Angle Analysis Method
Matching the two Roll Angles Balances the car

As shown many times in the past, our goal for most setups, especially on asphalt, is to match the desired roll angles of the front and rear suspensions. When we do this, the weight transfer becomes predictable and we can then have a balanced setup that will be fast and very consistent.

What Affects The Degree of Roll Angle? – The following are the effects that help determine what we can call roll stiffness, or the influence on the degree of roll angle for a suspension system. It is critical to know how these effects work before you ever begin to setup a race car.

103

Roll Center Height – The roll center for a suspension system is the bottom of the moment arm that tries to roll the car over when we are going through the turns. The length of the moment arm is a determining factor in the magnitude of the roll angle each suspension system desires. The longer the moment arm, the more the roll angle. This is a form of leverage where the longer the lever, the more power we have to move things, or roll the car.

So, if the roll center is higher, the moment arm is shorter, and the roll angle is less. If the roll center is lower, then the moment arm is longer and the roll angle is more. The roll center is a part of all suspension systems, AA-arm as well as the solid axle types and leaf spring types.

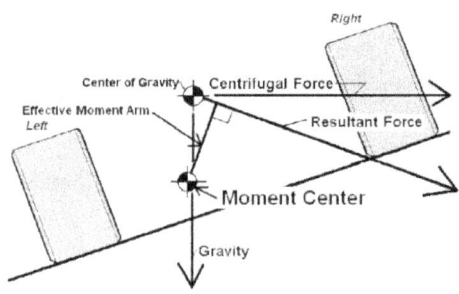

The Center of Gravity is the top of the Moment Arm and the Moment Center is the bottom of the moment arm. The height of the CG and the MC determine the length of the Moment arm and also the stiffness of the suspension in this AA-arm suspension. A longer MA causes a larger roll angle, and a shorter MA produces a smaller roll angle.

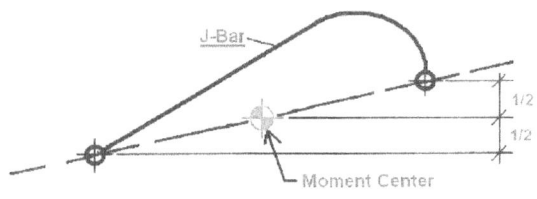

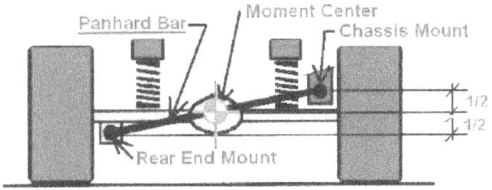

In a solid axle suspension, the Center of Gravity is the same, but the Moment Center is determined by the lateral restraint system, in this case a panhard bar. The MC height in this system is the average height of the center of the two end connecting bolts.

Center Of Gravity Height – The CG height of the sprung weight of the car, all of the components that are not included in the suspension, helps determine the magnitude of the roll angle. Like the roll center, the CG is a part of the moment arm and represents the top of the moment arm.

The higher the CG, the longer the moment arm and the greater the roll angle. The lower the CG, then the shorter the moment arm and the smaller the roll angle.

Spring Stiffness – Spring stiffness affects the roll angle. The stiffer the spring rates in a suspension, the less that system will roll. This is true if the spring rates are the same side to side (symmetric) or if they are not.

The spring stiffness is affected by the spring angle. In this installation, the rear spring being a coil-over design, is placed at a high angle off of vertical. This reduces the spring rate the car "feels" as well as shortening the spring base which is felt at the top of the springs in a straight axle suspension system.

This represents a swing arm spring installation. The coil-over assembly is mounted on the trailing arm and as such, creates a motion ratio. As the chassis moves vertically, the spring will move less than the chassis. The car feels less spring rate this way. The actual spring rate the chassis feels is a product of the motion ratio squared times the spring rate. In this car, that net rate is about half of the actual installed spring rate.

Spring Split – When there is a difference in spring rates side to side in a suspension system, that difference affects the roll angle. In a turn, if the inside (the turn) spring is softer than the outside spring rate, the roll angle is reduced. If the inside spring is stiffer than the outside spring rate, the roll angle is increased. These two statements consider that the overall average spring rate of the system is the same for both scenarios.

If we start out with two 200 ppi springs in a suspension system, and the roll angle is say 3.0 degrees, if we install a 175 ppi inside spring and a 225 ppi outside spring, then the overall average spring rate is the same at 200 ppi., but the roll angle will be less with the spring split, especially when the turn is banked with the inside lower than the outside.

If we again start out with two 200 ppi springs and then make a change to a 225 ppi inside spring and a 175 ppi outside spring, (the average spring rate of 200 ppi being maintained) then the roll angle will increase with the spring split.

Road racing cars must maintain a symmetry in spring rates, so they can never use this effect to tune the roll angles in the suspension systems on those cars. But for circle track cars, this is a major tuning tool.

Sway Bar Rate – For a suspension system that utilizes a sway bar type of mechanism, the spring rate, or stiffness, of the mechanism helps determine the roll angle. The stiffer the mechanism, the less the roll angle. The softer the mechanism, the greater the roll angle.

This shows a rear sway bar installation in a road racing car. Because this car is racing on a circuit where it must turn right and left, it cannot utilize spring split, or different spring rates side to side. Therefore, it must use a rear sway bar to help reduce the rear roll angle to match the front roll angle.

Lateral G-force – The lateral G-force helps determine the roll angle. The greater the G-force, the greater the roll angle. The lesser the G-force, the less the roll angle.

What is important to understand about the G-force influence is that for some suspension systems, when the G-force changes, the front and rear roll angles will change accordingly, but they don't necessarily change equally. If we put on new, faster tires, then the roll angle balance we had with old and slower tires can change. More on that later.

Track Banking Angle – The track banking angle influences the roll angle. The flatter the turn, the greater the roll angle becomes compared to a banked turn. So, the lower the banking angle, the more the roll angle. The greater the banking, the less the roll angle.

This is an easy concept to understand. For example, if we could drive around a turn with a 90 degree banking angle, the G-force would pull the car straight down into the track and there would be no rolling moment at all. So, the transition from a flat track to a banked one is taking us towards the 90 degree, zero roll angle, scenario.

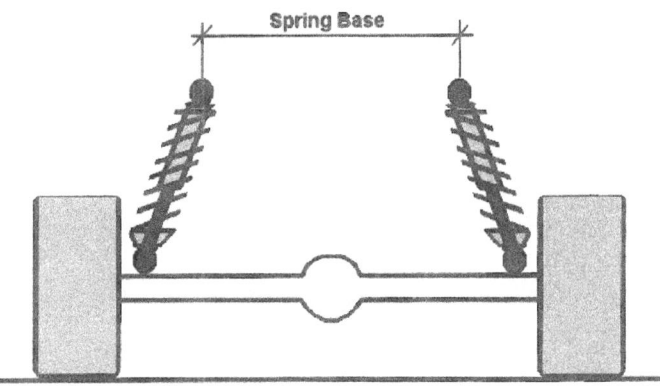

The banking angle has an influence on the roll angles for cars who run spring split, or different spring rates side to side on both the front and back. The lateral force that creates the G-force is combined with gravity and the resultant vector is pointed somewhere between the tires. This serves to create a smaller roll force and a mechanical downforce pulling the chassis down.

The rear springs on a straight axle rear suspension create the spring base. The car "feels" the spring base at the top of the springs. Some dirt cars have adjustable top mounts so that the upper mounts can be moved in or out. This creates an adjustment for the spring base width, which is a tuning tool used to change the rear roll angle.

Vehicle Track Width – The track width in a race car is known as the distance between the centers of the two tires at that end of the car. The wider this track is, the less the roll angle becomes for a AA-arm suspension system. A straight axle system is a little different and we'll talk about that next. In the AA-arm suspension, the spring rates are translated into wheel rates. The wheel rates become the suspension springs so to speak, and the wider the spring base, the less the roll angle. A narrow track is easy to roll whereas a wide track is harder to roll, just like any spring base.

Spring Base Width – As described above, the width of the spring base helps determine the magnitude of the roll angle. In a straight axle suspension system, the distance between the top of the springs represents the spring base. It is what the suspension feels as a spring base, similar to the AA-arm suspension feeling its spring base as the width between the tires.

And, in the same way as the AA-arm suspension, the wider the spring base, the less the roll angle and the narrower the spring base, the greater the roll angle.

This is an example of an adjustable right rear coil-over mount. It can be moved out to increase the spring base width for less roll angle, or in to shorten the spring base for more roll angle.

Why Match The Roll Angles? – We match the roll angles because we want the loads on the four tires to be ideal. That is, as we have discussed in earlier Lessons, at mid-turn, we want equally un-equal loading on the four tires. When this happens, the outside tires will carry the same load front to rear and the inside tires will carry the same load front to rear.

Remember in past Lessons where we explained that the most traction we get from a pair of tires on the same axle, or end of the car, is when they are equally loaded. Since in the real world, with weight transfer happening in the turns, there will never be equally loaded tires at mid-turn. The best we can hope for is to have as equally loaded tires as we can get and for the front and rear sets of tires to be equally un-equally loaded.

If the rear tires were more equally loaded than the front tires, then the rear would have more traction than the front causing a tight condition, or a car that understeers as the term implies. If the front tires were more equally loaded than the rear tires, then the car would be loose, or would oversteer as that term implies.

If the roll angles are not matched, then there will be uneven distribution of the loads, and the loading becomes very unpredictable and erratic. Un-equal loading and un-equal roll angles are the cause of slower mid-turn speeds and setups that don't perform the same as the laps increase on a continuous run.

Front With Higher Roll Angle – If the front suspension desired a greater roll angle than the rear, then it won't be able to achieve its desired roll angle and less weight will transfer at the front than would happen if the roll angles were balanced. Conversely, the rear would be pushed past its desired roll angle and then transfer more weight than would happen if the roll angles were matched.

In this case, the front would be more equally loaded and the rear more un-equally loaded. From what we have learned, the front would have more traction and the rear less traction. This car would be un-balanced and loose, or exhibit oversteer.

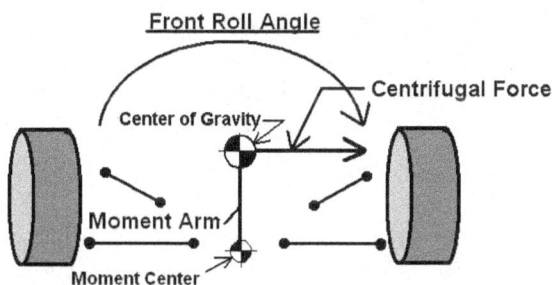

The Center of Gravity is the top of the Moment Arm for both the AA-arm suspension and the straight axle suspensions. The Moment Center is the bottom of the moment arm for both of these types of suspensions too. It is more difficult to change the height of the MC in a AA-arm suspension than it is to change it in a straight axle suspension. That is one reason why we usually tune the setup with the rear suspension spring rates and MC height for a stock car on circle tracks.

Rear With Higher Roll Angle - If the rear suspension desired a greater roll angle than the front, then it won't be able to achieve its desired roll angle and less weight will transfer at the rear than would happen if the roll angles were balanced. Conversely, the front would be pushed past its desired roll angle and then transfer more weight than would happen if the roll angles were matched.

In this case, the rear would be more equally loaded and the front more un-equally loaded. From what we have learned, the rear would have more traction and the front less traction. This car would be un-balanced and tight, or exhibit understeer.

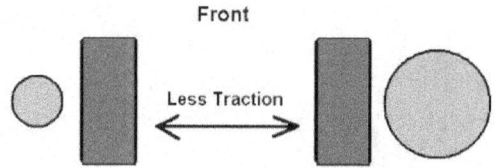

The result of an unbalanced setup where excess load transfers to the Right Front tire causing the rear tires to be more equally loaded. This would be a very tight handling car.

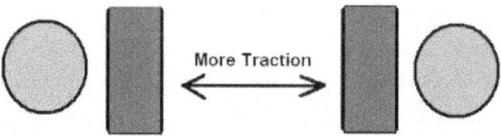

The sizes of the circles represent the amount of loading. If the car were setup tight as far as balance is concerned, then the rear tires will be more equally loaded and the front less equally loaded. From what we have learned about more equally loaded tires having more grip, this car would be tight. That is, the rear has more grip than the front.

Which End Do We Work On For Balance? – It is important to know which end to work on when we are trying to balance the setup. For a circle track car with a AA-arm front suspension and a straight axle rear suspension, we usually work with the rear suspension only.

That's not to say we cannot stiffen or soften the front spring rates, but if we have determined the correct stiffness of the front spring rates, it is much harder to effect a change at the front than at the rear. There are many more tools for adjustment at the rear in a straight axle suspension.

For a car with AA-arm suspensions front and rear, it becomes a matter of the platform attitude, or rake of the chassis when we decide which end to work with to achieve balance. For most applications of AA-arm front and rear systems, we adjust both ends if possible.

If the sway bar systems for one end is deficient, then we need to only work with that sway bar system to effect changes that will balance the setup. I have personally worked with a prototype car that from the manufacturer had a severely deficient rear sway bar system. The installed spring rates were fine.

The un-balanced condition this car suffered from was due to the fact that the rear sway bar was very weak and did not control the roll of the rear suspension. Through a series of tests, we determined that we needed more sway bar control and installed much stiffer components to solve the problem.

On some models of prototype cars, this one included, the sway bar is the primary tuning tool to create a balanced setup. We found a situation where the rear sway bar was not doing much work to help balance the car and the rear roll angle was much more than the front. When we redesigned the sway bar system and made it much stiffer in resistance to roll, the car became much more neutral in handling, and it came alive and picked up a lot of speed.

This is another car that when evaluated and balanced in roll angles became a multi-time winner at many different race tracks. This setup is more conventional and there are no tricks involved in setting it up. Once the balance is found, the setup is just maintained.

How Do We Work Towards Balanced Roll Angles? – If for example, we determine that the car is loose, or oversteering. This indication might come from the driver input, or tire temperatures. In all probability, the front is desiring to roll more than the rear. The basic way we would balance the roll angles and also tighten the car, or give more grip to the rear, is to do the following:

Decrease The Rear Spring Rates – If we decrease the average rear spring rates, we will then increase the rear roll angle. There is usually an optimum spring stiffness for a particular type of race car and we want to install spring rates that fall within that range of stiffness. So, this might not be the ideal solution.

For example, for a typical circle track car, the rear spring rates fall between 150 ppi and upwards of 300 ppi. We wouldn't think of installing a pair of 650 ppi springs to reduce our rear roll angle. It might work to do that, but the performance for the entry and exit portions of the track would be affected due to the extreme stiffness of those springs.

Introduce Rear Spring Split – Rear spring split has a pronounced effect on the roll angles. It is by far one of the most effective ways to change the balance on the race car. It is not a fine tuning tool, but a way to move quickly and in a major way towards a balance when the car way off. There are other ways to fine tune the balance that we will cover shortly.

Based on what we learned before in this Lesson, if we need to change the roll angles, we can introduce rear spring split. If the car is tight, with the rear out-rolling the front, we can soften the Left Rear (LR) spring rate, and the desired roll angle will become less. With this tight car, this change would serve to move more towards a balanced car. We can also stiffen the RR spring rate to increase the spring split to again reduce the roll angle and loosen the car.

If the car were loose, then if we increased the LR spring rate, or soften the RR spring rate, these two changes would increase the rear roll angle and tighten the car to bring it to a more balanced setup. Again, these are major changes and we will test our sample car in Part Two of this Lesson to see how much effect each of these changes produces.

For an example, for a circle track late model car running bump springs, we might split the rear springs and install a 150 ppi LR spring and a 275 ppi RR spring. That would produce a low roll angle. To equal that same roll angle with even rate rear springs, we would need to install two 650 ppi springs.

Rear Spring Rate Change – To a lesser extent, changing the average spring rate of the rear springs does change the rear roll angle. Stiffening the spring rates will reduce the roll angle and softening the spring rates will increase the roll angle.

As an example, if we stiffen the rear springs from a pair of 175 ppi springs to a pair of 200 ppi springs, we reduce the rear roll angle on a typical late model circle track race car by 0.2 degrees. That is not a big change. To make a comparison, that spring rate change is equals to a 0.25", or 1/4 inch change in the height of the panhard bar.

Changing The Height Of The Front or Rear Roll Center – Since the roll center is the bottom of the moment arm, we can make changes to raise or lower the roll center at each end of the car. Once again, for a AA-arm front and Straight axle rear combination of suspensions, the rear is much easier to make changes to than the front as far as roll center is concerned.

For a AA-arm front and rear type of race car, we probably wouldn't often make roll center height changes and would only work with spring stiffness and sway bar control.

For circle track cars, or for that matter, any car with the straight axle rear suspension, we can raise and lower the panhard bar. This device is called different things and goes by panhard bar, J-bar (because of its shape only), or track bar. Other devices that control the lateral movement of the rear end include a Watts link and the leaf spring.

Whatever system is used, and the panhard bar systems are by far the most popular and most used, we can more easily make changes to the rear roll center height by raising or lowering the bar, so that is what we do to fine tune the rear roll angle once we get close to the ideal angle.

An Example Of A Roll Center Height Change - The magnitude of the changes can be demonstrated easily. Using the setup software I described at the beginning of this Lesson, a typical late model circle track car might have a rear spring split of 150 LR and 275 ppi RR spring rates that are used to reduce the rear roll angle to match the front suspension stiffness. This car has a rear roll center height of 11.0 inches off the ground.

If we decrease the RR spring rate by 25 ppi to a 250 ppi spring, the reduced rate of the RR will increase the rear roll angle. To compensate for this change and maintain the same roll angle, we would need to raise the rear panhard bar (roll center) by 0.625 inches, or 5/8". As most veteran racers will tell you, that is a significant change in panhard bar height.

So, we can see where a seemingly insignificant 25 ppi change in spring rate at the rear makes a large change in the roll angle in the rear. In this case, the roll angle went up by 0.60 of a degree, or over a half degree. In Part Two of this Lesson, we will be installing springs and looking at the roll angles in typical types of race cars.

Spring Base Change – We can make changes to the rear roll angle by changing the width of the rear spring base. On some race cars mostly dirt cars, the top mounting point for the rear springs can be moved in or out. This changes the spring base width, and as we have learned, the spring base width is one determining factor in the final how large the roll angle is.

Again, for a typical late model race car, if we widen the rear spring base by moving the top of the two rear springs out by 2.0" each, this widens the spring base by a total of 4.0". This decreases the rear roll angle by a full half of a degree, or 0.50 degrees. This is also a significant change in the rear roll angle.

How Do Tire Loads And Forces Fit In? – In past lessons we talked about the importance of ideal tire loading and how to measure the spring forces to help achieve the ideal tire loading. When we can match the roll angles and balance the setup, we will then have ideal tire loading at mid-turn. One way to find the perfect balance, and matched roll angles, is to measure the existing spring force for a particular setup and then compare that to the spring force we would need to produce the ideal tire force.

So, not only do we have a tool to predict the roll angles and balance, we have a tool to measure the on-track spring forces that produce the tire loading. Once we know what is going on with the forces, we can then make adjustments to the setup to perfect the spring forces and ultimately the tire loading. We'll examine that process in Part Two of this Lesson.

Summary For Part One – This part one of Lesson Sixteen demonstrates how we can make changes to our setups to effect changes to the roll angles. We are working towards a balanced setup so that the wheel loads at mid-turn are ideal to produce the equal un-equally loaded tire loads that define a balanced setup.

For dirt racing, we don't necessarily match the roll angles in all cases. But knowing what changes will have what effect is important in any type of racing. We'll get more into the dirt setups in a separate section. For now, we come away with a better understanding about what we can change, and how much of a difference each change can make. There are major setup changes to get us close quickly, and more minor setup changes that serve to fine tune the setup.

Exam - In The Context Of This Lesson:

Which Part Of The Turn Do We Tune For First?

1) The entry
2) The mid-turn
3) The exit
4) All of the above

Mid-turn Steady State Is Defined As?

1) No acceleration
2) No braking
3) Maximum lateral loading
4) All of the above

Mid-turn Balance Means?

1) The car is not tight
2) The car is not loose
3) The front and rear roll angles are the same
4) The wheel loads are all the same

A Car That Is Setup Un-balanced And Tight Has?

1) Too much front roll angle
2) Too much rear roll angle
3) Equal roll angles
4) A low cross weight percent

We Can Measure Each Chassis Suspensions Roll Angles By?

1) Calculations using shock travels
2) Knowing the roll resistance of each end
3) Using a dedicated computer program
4) Using a pull down rig

For A Car With A Solid Axle Rear Suspension, We Tune Balance By?

1) Adjusting the front components
2) Adjusting the rear components
3) Changing the front geometry
4) Raising the rear of the car

We Can Adjust The Rear Roll Angle By?

1) Changing the spring rate split
2) Adjusting the rear roll center height
3) Changing the rear spring base width
4) All of the above

Which Rear Change Causes The Most Change In The Roll Angle?

1) Changing the average spring rate by 25 pounds
2) Changing the rear roll center height by ¼ inch
3) Introducing a 25 pound rear spring split
4) Changing the rear spring base by one inch

Race Car Technology – Level Three
Lesson Eighteen – Practical Applications – Part Two

This final Part Two of Practical Applications began in Lesson Sixteen will involve actually setting up race cars for the purpose of creating a dynamic balance. We won't be tuning with shocks, or for entry and exit performance here. This is all about mid-turn balance. We'll tune the other parts of the turns in Lessons to come. We need to get this process presented and understood by you before we go on.

Granted, there are a lot of race car types out there in the racing world and the students who enroll might have different interests and different race cars. The overall goal of Online Racing School with the Race Car Technology courses is to try our best to include information about all types of race cars.

The fact remains that circle track racing makes up the bulk of racing in the US. In other parts of the world, that might not be so true. We truly believe that the methods we present can be used, with some obvious tweaking, for every type of race car.

In this part Two, we will place a heavy emphasis on circle track race cars, only because there are a lot more of those here in the US where ORS is produced. But, it will be important to follow along as we go even if your racing is road racing stock cars, or formula and prototype racing. There is an interconnection between what we do for circle track cars and what would be done with other types of race cars.

The real difference between circle track cars and non-circle track ones is that it becomes easier to setup the circle track cars because they only turn one way. Race cars that turn both ways are somewhat limited in what can be done to achieve the balance we desire.

No matter what the type of race car, the latest technology in setup information works to get the car lower and more level to the race track. Even in the stock classes, teams are going ever softer on the ride springs to achieve a low attitude. This provides better aero efficiency as well as a much lower Center of Gravity.

How It's Always Been Done – Throughout racing history, the initial setup for a new car, or one we purchase used, but without information, is to engage in trial and error. If you can find the original manufacturer, or if your car is new and you are in contact with the car builder, they can provide a "ball park" setup to start you out.

If that is not an option, then most racer teams will seek out information from fellow racers who race in that division with that type of car. Or, they can ask consultants who may work alone or in partnership with some of the parts manufacturers like shock companies. After getting that first setup and trying it out, you're off and running into the setup world of racing.

This is the way it has always been done. We install a setup that might and might not include all of the unique characteristics of your car. Each car is different, and our sample car we use to develop the setups here is no doubt different than your car in many different ways.

To name a few, the CG is probably a little different due to different weight drivers, etc., the weight distribution, the motion ratios, the front and rear geometry, the bumps used if that applies, and more. No two cars are exactly alike even if they come off the assembly line together. Once you get the car, what you do with it makes it unique.

We're going to use very little trial and error to setup our sample car, and only then to fine tune the setup. We'll know all of the critical information about the car going in. When we get finished, we'll have the correct geometry front and rear, the correct weight distribution, and the correct spring rates. So, let's setup a race car shall we.

The stock classes as well as the Pro Late models are being setup in special ways. Experimentation by top teams throughout the last five years has yielded big gains in performance.

Choosing Spring Rates – Our choice of spring rates is dependent on the type of race car we will be racing and the actual type of setup we will run. In every class of race car, there are certain norms as far as spring rates.

For example, in road racing, both stock car and prototype, the spring stiffness is high and the front spring rates are usually 10-20% higher than the rear spring rates. These cars utilize front and rear sway bars to assist in controlling the roll of each suspension system. The overall control of the roll angles for road racing cars is tuned using the sway bars in most cases.

For asphalt circle track cars, there are presently two basic setups, conventional and bump setups. The rules for each division might allow or dis-allow the use of bump devices. The overall goal for most circle track race cars, dirt and asphalt, is to run lower in the front and particularly with asphalt cars, to reduce the overall average roll angle.

The combination of spring rates and sway bar rates serve to produce a lower attitude and one that has less roll angle. We've learned a lot about these various combinations over the past few years or so and we will pass that knowledge on to you. Now let's setup a real race car.

Choosing spring rates is an art that is backed up with science. We have tools available to us now that can predict and measure the chassis setup balance and the dynamic tire loading. There is less and less trial and error involved in creating modern day setups.

Conventional Metric Stock Car – The metric stock division race car based on a production car must use the stock control arms and stock rear suspension parts. The rear moment center is very high in these cars and can run from 13.5 up to 15.0 inches or more off the ground. This makes the rear of the car very stiff in roll resistance. So, we need to compensate for that high rear moment center by softening the spring setup in the rear. The roll angles will be higher than we see with the late models because the CG is much higher and the springs are softer overall.

A typical conventional setup for a metric four link stock car running a 1/3 mile track with 12 degree banking on 8" grooved tires might look like this:

CG @ 18.5" height.
G-Force = 1.25
Spring Rates: LF = 600 ppi, RF = 600 ppi, LR = 200 ppi, RR = 175 ppi, (this is a big spring front suspension)
Sway bar size = 7/8" solid,
Cross weight = 47.4%, (these cars have a high front percent, 52.5% in this case)
Rear Moment Center Height: 14.00".
Roll angles are 3.50 front and 3.60 rear.

This stock class car is built from a Metric Monte Carlo production car. These cars are very popular in the stock classes, but must be setup a little differently than other types of stock class cars that have.

Metric Stock Class Asphalt Running Big Springs

Corner Weights		Spring Rates		Rear Moment Center
738	648	600	600	14.0"
830	584	200	175	

Cross Wt % =   47.4%
Center of Gravity =   18.50
Sway Bar Rate =   0.875 w/ 0.187" wall
Roll Angles =   3.50 / 3.60
Rear Percent =   46.5%

When we refer to a "big spring" car, we car saying it uses stock dimensioned and designed lower control arms in the front with the springs mounted in the stock position. And, the rules might also require using the bigger 5" diameter springs at the rear also. Here we see the yellow springs in the front end of this stock chassis. The spring is resting in a pocket in the lower control arm and the top is controlled by a placement inside the frame rail. This spring can be adjusted for height by use of a screw jack bolt seen above the spring. When we are setting up the car, we need to keep in mind that the motion ratio of this big spring car is much different than if it were mounted like a coil-over in a late model.

Conventional Late Model – In days past, and again it is always good to know where we came from, the setups were what we now call, conventional. Even the conventional setups changed over the period from the late 1990's to the mid 2000's. A typical conventional setup for a late model stock car running a 1/3 mile track with 12 degree banking on 10" slicks might look like this:

CG @ 16.5.
G-Force = 1.50
Spring Rates at LF = 300 ppi, RF = 300 ppi, LR = 175 ppi, RR = 175 ppi,
Sway bar size = 7/8" solid,
Cross weight = 52.4%, (the low range of cross weight – it is common to use left side percent numbers)
Panhard bar was set at 9.00 left and 11.0 right.
Roll angles are 4.20 front and 4.20 rear.

Conventional Late Model Asphalt

Corner Weights		Spring Rates		Panhard Bar
758	616	300	300	9.00" / 11.00"
832	558	175	150	

Cross Wt % = 52.3%
Sway Bar Rate = 0.875" Solid Bar
Roll Angles = 4.20" / 4.20"
Rear Percent = 50.3
Center of Gravity = 16.5"

This is an actual setup I ran on a car in 1996 that won a lot of races in the mid-Atlantic region. It is balanced with the front and rear roll angles at 4.50 degrees.

In more conventional times just a few years ago, the race cars looked like this with the front valence riding several inches above the race surface. Now these cars ride much lower and flatter to the track. There is a difference of several tenths of a second in lap times between the two types of setups. We know because we ran head to head against top bump teams using conventional setups that were balanced.

In the mid-west, the cars ran on much flatter tracks for the most part, and the setup would change for the G-force, RR spring rate and panhard height. This car running on a 6-8 degree track using the same tires might make the following changes to the setup: RR spring to 150 ppi, G-force at 1.35 and panhard bar at 11.0" left and 12.0" right. Note that we had to move the panhard bar up 2.0" just to compensate for the softer RR spring.

Soft Conventional Late Model – Th pro, or limited, late model type of car is very similar to the super late model car in our next example. These are by far some of the most popular types of race cars in today's racing and in some cases these cars cannot run bump devices. So, we install soft front springs, larger sway bars and then tune the rear roll to match the front. A typical soft conventional setup for a late model car running a 1/3 mile track with 12 degree banking on 10" slicks might look like this:

Our sample car NOT running bumps will have the following setup:

CG @ 15.5".
G-Force = 1.75
Spring Rates at LF = 125 ppi, RF = 150 ppi, LR = 150 ppi, RR = 200 ppi,
Sway bar size = 1.500" with 0.25" walls, (these soft front spring setups need a much larger sway bar)
Cross weight = 52.8%,
Panhard bar set at 9.50" left and 10.25" right,
Roll angles are 2.50 front and 2.50 rear.

Soft Conventional Late Model Asphalt

Corner Weights		Spring Rates		Panhard Bar
738	648	125	150	9.50 / 10.25
830	584	150	200	

Cross Wt % = 52.8
Sway Bar Rate = 1.50" / 962 ppi
Roll Angles = 2.50 / 2.50
Rear Percent = 50.5

This chart represents a soft conventional setup where the car is designed to use coil-over springs in the front.

Soft Conventional Late Model Asphalt Big Springs

Corner Weights		Spring Rates		Panhard Bar
734	652	225	250	9.00" / 11.00"
834	580	150	200	

Cross Wt % = 53.1
Rear Percent = 50.5
CG = 15.5"
G-force = 1.75
Sway Bar Rate = 1.50" / 962 ppi
Track Banking = 12.0
Roll Angles = 2.30 / 2.30

Here we have a soft conventional setup where the car uses stock big springs in the front mounted in the stock location.

Bump Spring Late Model Setup – For the bump spring setup late model, we now get into the very stiff front spring rates creating much less roll angle. These setups are designed to get the car low in the front and to reduce the roll angle significantly. For the front spring rates, we add the ride spring rate to the bump spring rate. Also, these cars are usually allowed to run softer 10 inch slicks.

CG @ 14.0". (the nose is at least 2.0 inches lower than with a soft conventional setup)
G-Force = 2.00 (the lower CG and softer tires generate more G-force)
Spring Rates at LF = 1625 ppi, RF = 1625 ppi, LR = 150 ppi, RR = 275 ppi,
Sway bar size = 0.875" solid (we went back to a smaller bar)
Cross weight = 53.0% (there is a high range that some teams use by adding 6-7 percent)
Panhard bar set at 9.00" left and 8.00" right (a reverse split)
Roll angles are 1.10 front and 1.10 rear.

This is a Pro Late model, and part time super late model, that runs on bump springs. The setup sample car we listed was based on this very car that we setup several years ago.

NOTE: Bump stop setups are very similar to the bump spring setups. We only need to determine the spring rate of the bump at the compression travel it is operating at. If we record a gain of 750 pounds of force in a half inch, then we can assume the bump spring rate is double that in one inch, or 1500 ppi. We then add that rate to the ride spring rate.

Bump Late Model Zero Roll Setup (Experimental) – For the bump spring setups used today in the super late model division, we are seeing some different setups. This design was first spoken about in a Circle Track article I wrote some years ago. Everyone at that time was trying to reduce the roll of the cars. So, I spoke about creating a zero roll angle setup and how to do it. Now we see some teams, at some race tracks, trying that concept.

The design involves creating a positive roll angle at the front and a negative roll angle at the back with the same roll angle number, but in reverse. The front wants to roll right, and the rear wants to roll left. These two roll angle desires cancel each other out leaving us with a net zero average chassis roll angle.

This school is about teaching and that is why we are making this known. This is a concept that is used in racing today. We don't necessarily promote it, just acknowledge it exists. Here is how it works.

For our bump super late model sample car, we generated a front roll angle of 1.10 degrees. If we arrange the rear spring rates (using rear spring split) and raise the rear moment center height (panhard bar height), we can realize a negative 1.10 degrees of roll angle. That negative 1.10 degrees of roll will cancel out the front positive 1.10 degrees of roll angle.

Here are the numbers:

CG @ 14.0".
G-Force = 2.00
Spring Rates at LF = 1625 ppi, RF = 1625 ppi, LR = 150 ppi, RR = 300 ppi, (increased rear spring split)
Sway bar size = 0.875" solid
Cross weight = 59.0%, (the high cross weight puts load on the LR tire compensating for a loose setup)
Panhard bar set at 11.50" left and 9.00" right, (raised the panhard bar to reduce the rear roll angle)
Roll angles are 1.10 front and (-) 1.10 rear. (now we have a net zero average chassis roll angle)

If you look at the front valence on this car, it appears that the setup is based on the zero roll angle concept that we explained. We call this the equal and opposite roll angle setup. This is becoming popular and testing will tell if it is better than a setup that has equal positive roll angles.

Late Model Asphalt Bump Setup
Zero Roll Angle

Corner Weights		Spring Rates 125/1500 150/1500 1625 1650		Panhard Bar
750	624			11.0" / 9.0"
841	550	150	300	

Cross Wt % = 53.0
CG Height = 14.0"
Rear Percent = 50.5%
Sway Bar Rate = 1.50" / 0.25" wall
Roll Angles = 0.92 / (-) 0.99
G-force = 2.00
Track Banking Angle = 12.0

This is the basic zero roll angle setup. This setup uses a stiff RR spring to help achieve the negative rear roll angle. There are other ways to setup this zero roll concept.

Late Model Asphalt Bump Setup
Loose In With RR Pre-load

Corner Weights		Spring Rates 125/1800 150/1800 1925 1950		Panhard Bar
732	654			13.0" / 12.0"
836	578	150	150 / 225	

Cross Wt % = 53.2
Rear Percent = 50.5%
Sway Bar Rate = 0.875" / 135#
Roll Angles = 0.85 / (-) 1.20
Bump Rate = 1800 / 1800

If we install a RR spring rate of 150 and pre-load it with a force equal to the previous stiffer 300 ppi spring to support the RR during entry and at steady state mid-turn, then we can create a setup that is loose into the corner. Because the rear roll is a higher negative number than the positive front roll angle, the front will have more grip. This makes the car loose in, or turn in much better. At mid-turn, as the car begins to accelerate, the added load being transferred from front to rear loads the RR spring beyond the pre-load and now engages the softer 150 ppi spring. As we will see, this softer spring rate tightens the car off the corner.

Late Model Asphalt Bump Setup
Tight Off With RR 150 Spring

Corner Weights		Spring Rates 125/1800 150/1800 1925 1950		Panhard Bar
732	654			13.0" / 12.0"
836	578	150	150 / 225	

Cross Wt % = 53.2
Rear Percent = 50.5%
Sway Bar Rate = 0.875" / 135#
Roll Angles = 0.85 / + 1.20
Bump Rate = 1800 / 1800

Once the load on the RR spring surpasses what is needed at mid-turn, the 150 ppi spring rte comes into play and we can see where the rear roll angle goes from a negative (-) 1.20 to a positive 1.20. Since the rear roll angle is now more than the front roll angle, the setup becomes a tight setup helping to drive the car off the corner.

We aren't sure this is the best setup you can run, but we acknowledge it is being used. In theory, it might work very well. What it also does is try to twist the chassis. A balanced setup is where both ends of the car are trying to roll to the same angle. In this setup, the two ends are not in sync and are working against each other.

Road Racing The Late Model – For road racing using a late model type of chassis, we are turning right and left, so this setup must be different than what we develop for a circle track car. We cannot use spring split in the rear to help reduce the rear roll angle. Therefore, we will need to use a rear sway bar.

There are several other differences we see. The overall spring stiffness is going to be much more for the road racing car. And, the weight distribution must be such that we end up with a 50/50 cross weight percent. The cross weight must be the same for right and left turns.

The weight distribution is a very critical element in these cars. If possible, we need to find a front to rear percent that works to create an ideal weight distribution at mid-turn when we start out with a 50% cross weight. We think we can. Here is a typical late model car that has been converted to road racing. We use the same basic car as in the previous circle track examples, only we make the suspensions symmetrical side to side.

CG @ 15.5".
G-Force = 1.30
Spring Rates at LF = 600 ppi, RF = 600 ppi, LR = 400 ppi, RR = 400 ppi
Front Sway Bar Size = 1.375" w/ 0.125" wall
Rear Sway Bar Size = 0.75" solid bar
Cross weight = 50.0%, (the cross weight percent must be 50% for a road racing car)
Panhard bar set at 10.00" left and 10.00" right,
Roll angles are 1.54 front and 1.63 rear.

This basic setup was run on an AGT class Grand Am road racing car that won three championships in the early 2000's.

Road Racing A Late Model Asphalt

Corner Weights		Spring Rates		Panhard Bar
666	666	600	600	10.0 / 10.0
634	634	400	400	

Cross Wt % = 50.00
Front Sway Bar Rate = 1.375" / 454#
Rear Sway Bar Rate = 0.750" / 60.0#
Roll Angles = 1.55 / 1.55
Rear Percent = 48.80

This is what a circle track car that has been converted to a road racing car might look like. The cars have more roll angle than a circle track car due to the fact that they must turn both ways, so we cannot utilize spring split or bump setups in these cars. Roll angle is not a bad thing if it is controlled.

In a circle track car that has been converted to a road racing car, or any AA-arm front and solid axle rear road racing car, it is a necessity to run a rear sway bar to help control the rear roll. Most of these sway bars are small in diameter being 0.75" or smaller. This one is adjustable for arm length.

Summary - How we finalize the setups we run beyond spring rates and roll center heights all depends on the goals for rear steer and the shock package that gets us into and off of the corners. In the coming Lessons we will explain how to select shocks and how to tune your setup at the race track. For now, just know that we all need a starting point that represents the basic, baseline balanced setup.

Keep in mind that any setup you get from someone else will not be what they win with. Nobody gives out the winning setup. It will be generic. Even coming from the chassis manufacturer, the setups you might get will be generic and the best setups might be saved for the house car, or a favorite race team that shows the manufacturers product well and that has already has some success.

The best setups are developed inhouse and are best kept a secret. As we stated, every race car is a little different than the next, even if they come off the assembly line together. Concentrate on creating an ideal setup for the chassis you run that provides performance based on the dimensions and weights of that chassis, not someone else's.

Next, we will discuss how shocks can enhance the entry and exit performance now that we know how to develop the setup that created a balanced mid-turn condition. A few Lessons later, we will tell you how to fine tune the setup at the race track.

Exam - In The Context Of This Lesson:

The Primary Part Of The Turns We First Setup For Is?

1) The entry.
2) The mid-turn
3) The exit
4) All of the above

The Primary Setup Goal For All Types Of Race Cars Is?

1) A balanced setup with equal roll angles
2) Soft front springs
3) Higher rear moment centers
4) Very little roll angle

The Metric Stock Class Car Has What Unique Characteristic?

1) They un big springs in the stock location
2) The C is higher
3) The rar moment center is very high
4) All of he above

The Transition From Conventional To Modern Setups Involve Which?

1) The use of soft front ride springs
2) Use of high rear spring split
3) The use of large sway bars or stiff bump devices
4) All of the above

Road Racing Cars Differ From Circle Track Cars How?

1) They run much stiffer spring rates
2) They run rear sway bars
3) The cross weight must be 50%
4) All of the above

Race Car Technology – Level Three
Lesson Nineteen – Advanced Racing Shock Tuning In Transitions

In keeping with the theme of the setup part of the Online Racing School, Race Car Technology courses, we tune the Mid-turn portion of the turns first. Then we tune the Entry and then Exit in that order. One of the major tools we use to tune entry and exit are the shocks. With just a basic understanding of how the dynamics of the race car works, we can choose our shock rates correctly to complement our setups.

In the modern world of racing, for both dirt and asphalt competition in circle track racing, and also for road racing, shocks have become one of the most important fine tuning tools we have. They can complement the current setups, especially the bump setups, if applied correctly.

The following information is useful whether you are running more conventional setups, bump setups or road racing. Shocks affect the speed of movement and the load distribution at the four corners of the car as we transition from high speed to minimum steady state mid-turn speed, and then to up high speed again.

One of the most basic and important things to remember is this. Shocks do not affect the handling of the car if they are not in motion. To say that you can tune your mid-turn, steady state, handling with shock changes is wrong. Any competent shock expert will tell you that up front.

Even though we cannot affect the mid-turn handling with shocks, conditions that begin with corner entry that negatively affect the mid-turn handling can be solved with shocks. This is where the thought that shocks affect mid-turn handling began. If there are no entry problems and you need to solve mid-turn handling problems, shocks are not the answer.

Shocks Work With Springs - Controlling wheel movement would be much easier if all we had to work with was the shocks. But in reality, our race cars are supported by a set of springs. Basically, we always want to match our shock rates to the spring rates, and/or the bump device you might be using.

We can influence the load distribution on the four tires through the use of different rates of shock resistance to movement. For some types of race car, the entry and exit performance can be tuned by the use of shocks. We must remember that shocks do not affect loading on the tires when they are not in motion. Shocks only influence loads when they, and the suspension, are in motion.

For certain types of race cars, it becomes important to allow the corners of the car that are compressed to rebound and stay in contact with the track. The car on the right is allowing the left front tire to remain in contact with the track surface. The car on the left has the left front tire riding off the track, maybe as a result of too much rebound in the LF shock.

Springs resist compression and promote rebound. The shock doesn't need to do much compression control because the spring resists compression. It is hard to compress a spring, and the more rate the spring has, the harder it is to compress.

Springs promote rebound. When a spring is compressed, it wants to return to its normal free height. The more rate the spring has, the faster it wants to return to its free height. Less than adequate shock rebound control causes the suspension to move too quickly.

So, based on the above, in all situations, the shocks rebound rate will always be greater than the compression rate for all suspension shocks because the spring helps resist compression and promotes rebound.

As we install stiffer springs, we would naturally need to increase the rebound resistance and decrease the compression resistance. Stiffer springs would include adding bumps to one or more corners of the car. Each bump device has a spring rate that is measured in the range of motion it is operating in. Bump springs have a more consistent rate than bump stops and the rates of both are added to the equivalent ride spring rate.

With bump setups, we may say we have a soft spring setup because we install 150 ppi ride springs at the front in our coil-over car, but when we add the stiff bump rates to the ride spring rate, we end up with 1500-2000 ppi or more for the front spring rates.

Therefore, with those setups, keeping with the idea that the shocks must control the installed spring rate, we must run shocks with a rebound rate upwards of 1500-2000 pounds at 3-5 inches per second of speed.

The shock must control both the ride spring and if so equipped, the bumps spring or other bump device. The shocks control the overall spring rate and that includes all of the springs.

"Tie-Down" Shock Terminology – This brings us to an important discussion. The use of the term "tie-down" has been around for some time to describe shocks that are high in rebound resistance. The idea initially was that if we install these high rebound shocks, we can tie one or more corners of the car down and keep the tire "attached" to the track. This way of thinking is completely wrong.

First of all, none of the tires are connected to the track surface, and they are free to rise and fall according to what the other three corners of the car are doing at that instant on the track. If our setup causes that corner to move vertically because load has shifted off of it, then the load will come off regardless of how much "tie down" we design into the shocks. You just won't see much suspension movement if the rebound is too high.

We may be led into thinking that the "tie down" worked. No, it did not. If enough load comes off that corner, or the setup is unbalanced enough, the tire might well come off the track surface even without suspension movement. The existence of lack of movement doesn't mean the loads on the tire stayed the same.

This unloading of the tire will occur on the inside front or inside rear tire normally in a race car while turning. So, we cannot tie the inside front tire down and we cannot tie the inside rear corner down, although I've seen some teams try. Lack of suspension movement does not mean the load remains on the tire.

I repeat: Lack of suspension movement does not mean the load remains on the tire.

What we can do is control the spring rate of the ride spring plus any bumps we have installed. If your left rear spring is a 175 ppi spring, then you need to use a normal shock rebound rate associated with that spring or risk taking much of the load off that tire on entry into the corners. On some tracks when using too much rebound, not only will you be loose in, but also loose off due to the way the track banking transitions.

For race cars that ride on very stiff springs, the motion of the chassis is minimal. If the shocks offer control only when they are moving, then with little motion, there will be very little control offered by the shocks. Instead of moving these shocks through inches of motion on the shock dyno, they should be cycled through increments of one inch at the most in many cases.

Entry Tuning With Front Shocks – Let us understand one thing up front in our discussion about shock tuning. We have made it very clear that if the shock is not moving, it has no control over load distribution. When using bump setups where the front suspension moves onto, and rides on, the bumps on initial corner entry, there will be very little movement of the shocks throughout the turn until the car exits the corner and load shifts off the front causing the shocks to come off the bumps.

So, in relation to bump setups, the entry discussion presented here is mostly irrelevant. This discussion is for setups and race cars using more conventional springs to support the front suspension.

Moving on: If we split the front shock compression rates with a RF shock using a stiffer compression rate than the LF shock, then, while the suspension is in motion due to load being transferred to the front on entry, the RF suspension will move slower than the LF suspension. Additional load will be transferred onto the RF and LR tires causing a momentary increase in the cross weight percent in the car.

It is important to note here that the load transfers almost immediately when a force is presented to enact that transfer, such as applying the brakes. If on entry we transfer 300 pounds from the rear to the front, the 300 pounds goes to the front in an instant.

The distribution of that 300 pounds between the two front tires (while the suspension is in motion, and while in the process of assuming a new attitude that will support the additional load) will depend entirely on differences in stiffness of the suspension systems at all four corners. Stiffness is defined as the resistance to movement influenced by the shocks and springs.

The result of all of the above is this. The slower moving (or stiffer) corner in compression will momentarily retain more of the transferred load while the suspension is in motion.

Cross weight, as we commonly know it, is defined as the percent of the combined RF and LR weight divided by the total vehicle weight. If the cross weight percent increases, then a car turning left will be tighter on entry and the car might be faster if that is the desired effect. This is exactly why it has been said that a stiffer RF shock will speed up load transfer to that corner, although the reality of that statement is not entirely true.

Later in the entry event, some of the load that has been transferred onto the RF due to that corner being stiffer in compression will transfer over to the LF tire as the shocks stop moving and the car reaches a steady state at mid-turn. Then the normal cross weight you set the car at will apply.

If the car is already tight on entry, after having eliminated common causes of tight entry such as rear miss-alignment, rear steer or brake bias issues, then an opposite effect can be utilized. If we install a LF shock that is stiffer than the RF shock, and/or run a stiffer LF spring than on the RF, then we can effectively reduce the cross weight in the car on entry while the suspension is in transition by loading the opposite diagonal, the LF and RR. As one diagonal goes up in percentage of supported weight the other goes down.

The front shocks on bump type of setups result in very little movement from the point of initial entry until the car exits the corner and begins acceleration. This lack of movement means that the front shocks have very little influence on entry characteristics.

Entry Tuning With Rear Shocks – In a typical circle track car turning left, the left rear corner is usually sprung soft to enable the high spring split across the rear to reduce chassis roll. To keep from running a very stiff right rear spring rate, we use a lighter left rear spring in the range of 125 to 150 pound per inch.

Now we need to setup our shocks to work with that light spring. Since our spring rate is low, our shock rebound only needs to be sufficient to control that softer spring rate.

Most left rear shocks are valve'd to a higher rebound setting than what is needed because back in earlier days using conventional setups, the LR spring rate was higher. If we use a shock that worked for the older setup and is too high in rebound, then on entry to the corner, tire loading will be reduced on that tire and the car will become loose.

This is a very common problem and easily solved by just reducing the rebound in the LR shock. The LR shock does not need very much rebound, only enough to control the spring that is installed, or less.

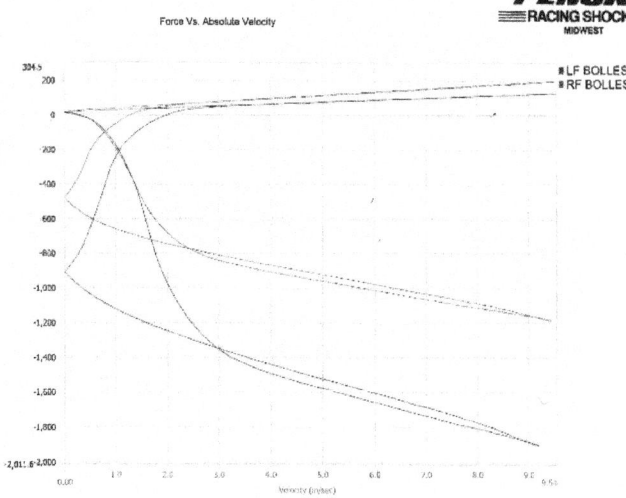

Here we see a more typical pair of front shocks built for bump setups. The LF shock (shown as a blue line) has a "nose" rate (initial force at the start of movement) of around 900 pounds, and around 500 pounds for the RF shock (red line). The rate at 3.0 ips then climbs to 800 for the RF shock and around 1,350 pounds for the LF shock. These relatively high rates are necessary to control the combined spring rates of the ride spring plus the bump device.

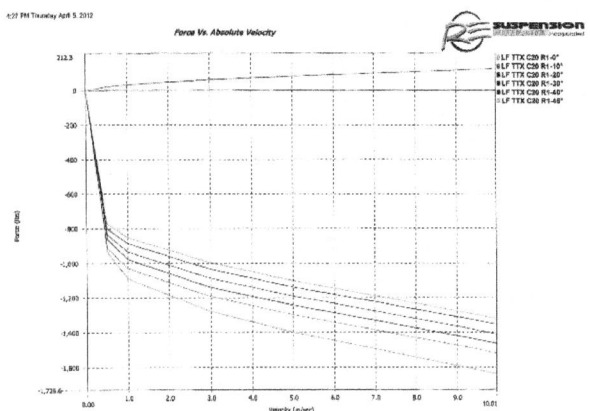

This represents a shock dyno readout sheet showing the various shock rates through different bleed settings. The bottom readings are for rebound. We can see that this shock is adjustable for rebound from 1,000 pounds of resistance to almost 1,300 pounds of resistance at 3.0 inches per second speed of movement. This is also a digressive shock in that the resistance initially ramps up when approaching 1.0 ips speed and then tapers off as the speed increases up to 10.0 ips of speed.

It is useful to use shocks that are adjustable for rebound and compression so that when you change springs or bumps, you can then adjust the shocks to match the new rates. This shock has a rebound adjuster that we can see just above the eye. A tool is used to "sweep" the adjuster to restrict or open up a needle valve in the shock shaft.

Exit Tuning Using Split Valve Shocks — Corner exit performance can be improved by utilizing the shocks. This is done by either working with the compression settings in the rear shocks and/or working with the rebound settings in the front shocks. In a car with equal rear spring rates, a stiffer compression setting in the LR shock than in the RR shock will load the LR and RF corners while the shocks are in motion.

Load is transferred to the rear under acceleration and while the rear suspension is in motion, this split will tighten the car by increasing the cross weight percent. Keep in mind that this motion and event happens very quickly, so the effect is short termed. A shock with a stiffer rebound rate at the LF corner can help accomplish the same effect by causing a slower movement of that suspension and a more rapid transfer of load off of that corner which in turn increases the percentage of load supported by the RF and LR tires.

If the RR spring is somewhat stiffer than the LR spring, there will be a loosening effect on acceleration when load transfers to the rear. The stiffer RR spring causes more of the load to be placed on the RR and LF corners reducing the cross weight percent. This event last much longer than the event caused by the shock being in motion.

By installing a LR shock with a much stiffer compression rate, and a RR shock with a much softer compression rate, you can momentarily nullify the negative effect of the stiffer RR spring. This serves to equalize the unequal resistance to compression due to the dissimilar spring rates and helps keep the car tight on exit. Again, keep in mind that this motion and event happens very quickly, so the effect is short termed.

A more permanent cure for this problem is to run a softer RR spring that has been preloaded to a force equal to what is needed to support the RR corner at, or near, the mid-turn portion. Then when the load has been transferred to the rear on acceleration, the RR will compress at the softer spring rate. In this way the load is more evenly distributed among the LR and RR tires. If you understand this statement, you have found a nugget of gold to use in setting up your race car.

This car has very little rebound in the front shocks as evidenced by the high attitude of the front valence when accelerating off the corner. With this little control, the springs in this bump setup are not being controlled properly through the entry and mid-turn portions and that can cause problems on a track that is rough.

Putting All Of This to Good Use — In order to utilize the configurations we have discussed here, we must be able to use a range of different rates of shocks in order to find the right combination for our car at a particular race track for a particular setup. For a team that races at only one track, the process is fairly simple.

You would experiment to find the fastest set of shocks and ones that suit the driver's style and stick to those while also staying within the boundaries of physics. For teams that travel to different tracks, some changes might be necessary if the setup (read as spring rates) needs to change and/or the track layout is different from the track you are used to.

Most shock experts agree with these basic statement:

1) The shock package should be softer overall when racing on dirt and when the track is flatter when on asphalt for the conventional setups.

2) Get your basic setup close to being balanced by tuning for mid-turn before trying to tune with shocks. Shocks cannot solve basic handling balance problems that mostly occur at mid-turn.

3) Higher banked tracks require a higher overall rate of shocks and springs as opposed to flat tracks. This is because of the higher speeds and the extreme amount of downforce.

4) Shocks that are mounted farther from the ball joint should have more overall compression and rebound control than if they were mounted closer to the ball joint. That is because with each inch of travel of the wheel, the shocks that are mounted farther away will move at a slower speed which means less resistance in both rebound and compression. (See the shock graphs)

This is also true of shocks mounted at high angles to the direction of motion.

5) Of the two transitions, tune entry performance first. If there are no entry problems, make small changes if you want to experiment to see if entry speeds can be improved. Entry problems include a tight or a loose car. By far the worst problem would be the loose-in condition. This can involve an alignment problem, but far too many times, the problem involves a LR shock that is too stiff in rebound.

6) Tune exit performance last. Exit problems can include a car that pushes under acceleration or one that goes loose under power. Be sure that you do not have a Tight / Loose condition where the car is basically tight in the middle and goes loose just past mid-turn. This is fixed with spring rate and/or panhard bar adjustment, etc. and is the first thing we are supposed to work on.

7) On dirt race tracks, reduce rebound settings on the left side and decrease the compression rates on the right side for dry slick surfaces to promote more chassis movement. This helps to maintain grip as the car goes through the transitional phases of entry and exit.

8) For the bump setups on asphalt, the whole shock package must be much different than when running conventional or soft conventional setups. The bump spring rates (either bump rubbers, bump stops, or bump springs) will be very high and so the shock rebound rates must match those high rates in order to control the ride spring and bump device.

Using a spring/bump combination that is rated in the 1500-2000 ppi range will need a shock that is rated at around near that range in the 3 - 5 inches per second range of movement. Usually the ultra-low speed rate of the shock will be comparatively high too and it is helpful to use a "nose" rate of between 500 and 800 ppi or more at zero inches per second speed of movement.

Nose means that it takes that much force to cause the shock to move initially. The shock will not move until that amount of force is applied to the shock.

The angle of the shocks as well as the motion ratio to wheel movement both affect the rate of the shock that influences wheel movement and chassis movement. A high angle like this one causes a reduction in efficiency of almost 20%. If this shock were rated at 150 pounds at 3.0 ips, then when installed on the car would only provide around 120 pounds of rate at the same vertical wheel speed. With this installation, 3.0 ips of wheel movement would equal 2.4 ips of shock movement. The same is true for motion ratio at the front end.

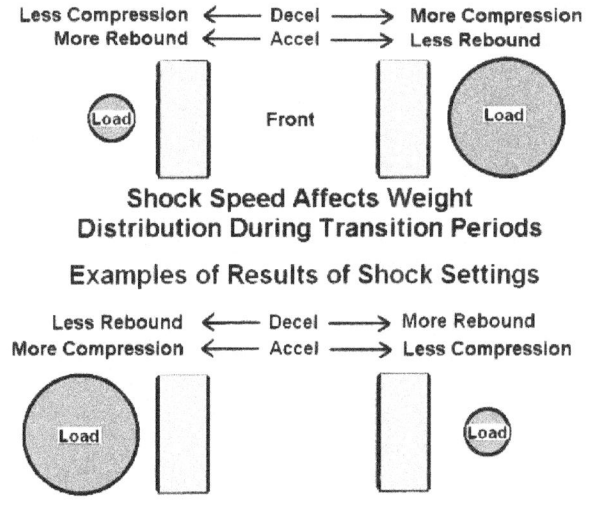

This chart represents how a shock can provide added loading on the RF and LR tires during the transitions listed. This is true for setups where the shocks actually move during these transitions and the loading influence only occurs for a short period of time, while the shock is actually moving.

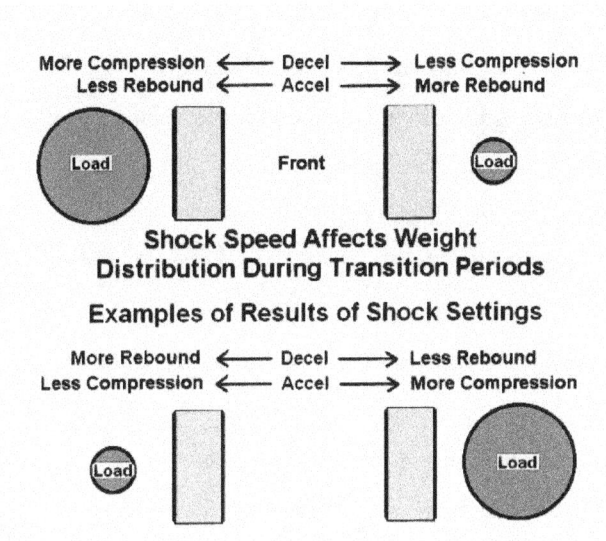

This chart represents how shock tuning can influence the added loading on the LF and RR tires during the transitions listed. Again, this effect is only present for a short period of time and only while the shock is in motion.

It is recommended that most amateur race teams consult with shock retailers and/or consultants that specialize in shock technology. If you don't have you own shock dyno to evaluate your shock rates, then seeking outside help is a must. Then the team can make intelligent decisions on how to setup the shock package.

Summary – The shock rates and changes provided here are representative of trends that can enhance your handling package. Before any of this can work, the setup must be balanced, the steering characteristics must be ideal and the car must be aligned properly. If not, nothing you will do by tuning your shock package will fix those problems.

Shock tuning is the last thing to experiment with in order to try to increase the race cars performance. It is a necessary step in finding the ideal total handling package. That said, before setting up your car and choosing your shocks, evaluate what you will need by matching the shock rates to the spring rates you will run.

Exam - In The Context Of This Lesson:

Shocks Affect Which?
1) Mid-turn handliing
2) The speed of chassis movement
3) How fast weight transfers
4) The amount of weight transfer

Shocks Do Their Work When?
1) They are mounted closest to the ball joint
2) The car is at mid-turn
3) They are in motion
4) They are mounted near the springs

Shocks Control What?
1) The compression of the springs
2) The rebound of the springs
3) The chassis attitude during transitions
4) All of the above

In Most Cases, Shocks Are Designed With?
1) More rebound than compression control
2) More compression than rebound control

Using A Tie-down Shock Is A Way Of?
1) Keeping the tire in contact with the track surface
2) Maintaining loading on the tires
3) Controlling chassis dive
4) None of the above

Loose-In Condition Can Be Cured By Which?
1) More compression control in the LF
2) More rebound in the LR
3) More compression in the RF
4) More compression in the RR

Loose-Off Condition Can Be Helped By Which?
1) More compression control in the LR
2) More rebound in the RF
3) More compression in the RR
4) More compression in the LF

Bump Setups Require More What?
1) Compression in the front shocks
2) Rebound in the rear shocks
3) Front rebound to control the high spring rate
4) Front compression to control the high spring rate

Which Do We Tune First?
1) Entry handlling
2) Exit handling

Race Car Technology – Level Three
Lesson Twenty – Testing and Tuning The Setup At The Track

To this point in RCT Level Three we have studied the important concept of balance, forces, roll angles, weight transfer, dynamic loading, angle of attack, what makes traction and force verses weight. We learned how a sway bar can be used as a spring. Then we ran through various setups for the mid-turn performance and the use of shocks for the transitions. All of that leads us to this all-important lesson.

Now we go to the track and try all of this out. One of the primary attractions of racing is the fact that everything you do with your race car will be evident once it hits the track. If it is fast, you did your job well. If not, there is more work to do, period. Every setup has its day on the track.

Here we will explain how to do a test session and how to tune the setup if you think it needs a bit of improvement. Hopefully this takes place before the season starts and there is enough time to properly examine how your setup is working.

On asphalt, the setup we end up with is probably the one we will qualify and race with given small changes between the two if the rules allow. For dirt, we will make setup changes that may be required for the different track conditions. That does not mean we cannot test on a track that is consistent.

A primary goal for dirt cars might be to just learn the process of making changes to meet the track conditions. There is an order and logic to adapting to changing track surface grip levels. Becoming comfortable with making those changes can be a huge performance gain.

The following is a description for testing a circle track late model type of car. No matter what type of race car you will be using, many of the important points will pertain to your car. Obviously, for road racing, there will be no spring split, weight distribution tuning, or other circle track specific changes. We will address road racing with stock cars and formula cars because we have experience with those types of cars.

Once we get the setup established, we need to go test at the race track. Not only do we need to evaluate the setup for balance and speed, we need to compare our performance to other cars we will be racing against. We never need to seek the fastest lap times without regard for creating the dynamic balance we have talked about continually in this school. Some setups are fast for a few laps, but in the long run, the fastest lap times at the end of the race are what wins races.

Road racing cars are more limited in what can be changed to create the balanced setup. The springs and suspension must remain symmetric and the weight distribution must stay at 50/50% cross weight. We tune these types of cars with spring rates and sway bar stiffness mostly. If the car has a solid axle rear suspension, we may be able to change the rear roll center height if it is equipped with an adjustable panhard bar.

Goals - The overall goal of testing is to develop the setup combination that will be both initially fast and also stay fast for a long time. It should be good on the tires, comfortable for the driver, and out run the competition all the way to the last lap.

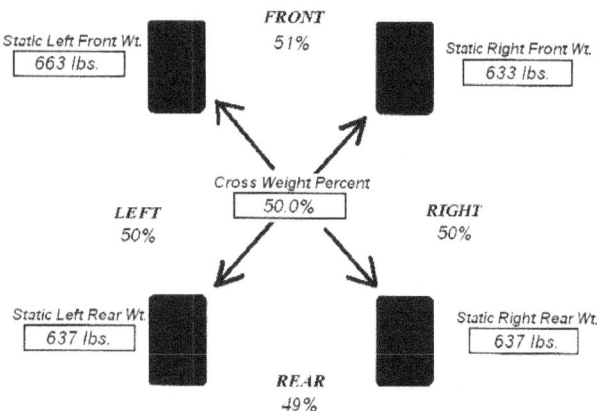

The road racing cars cannot adjust the weight distribution as far as cross weight is concerned. They may be able to move weight fore and aft to arrive at a weight distribution that works with a 50-50% cross weight. This is one of the secrets to creating the perfect road racing setup.

Cross Weight Related to Front Percent

		Front Percent		
46%	48%	50%	52%	54%
59.1% X	55.6% X	52.1% X	48.6% X	45.0% X
54%	52%	50%	48%	46%
		Rear Percent		

This chart shows the relationship of cross weight to front-to-rear percent. In a typical circle track race car, or even a circle track car that has been converted to road racing, there is an ideal cross weight that will work to provide the ideal wheel weights at mid-turn, based on the front to rear percent of weight distribution. Based on this, a road racing car would need to design the car with around 48.8% rear weight distribution. That way, the car can have a 50% cross weight distribution that will work for turning left and right.

Pre-testing Preparation / Planning – It is most important to know your car before you go to the race track for practice or testing. All of the things we talked about in Levels One through Three are what leads up to the setup. If those items are not in order, there is little reason to go test.

If you are going to a new track, take into consideration the banking and transitions. If it has a different banking angle than you are used to, a different spring setup might be in order. High banked tracks need higher spring rates overall and have little need for traction enhancing technology. If the track is flatter, include methods of creating bite off the corners into your planning.

Collect information before you go to your test. Check Ackermann, check rear alignment, eliminate bump steer and make rear steer settings that are appropriate. Weigh the car with different spring combinations and note what changes need to be made to bring the original weight distribution back to the baseline.

You should make notes detailing what needs to be done to the car before you go out onto the track. Also, list your test session priorities and post them on the car. This team was preparing the car before a big test and wanted to make sure all of the items had been taken care of and checked before departure. Once at the track, we tend to forget the order of testing we discussed at the shop.

Arrival – On arrival at the track, establish a pitting position for the car that is relatively level. You should have easy access to the tool cart as well as the trailer and other track facilities that may be needed. Mark the spots around the tires so you can always park the car in the same position after each run.

Weigh the car before testing and after all of the testing is done at the end of the day, re-weigh the car to see how the weight distribution might have changed from the various adjustments. If the track scales are different from yours as to cross weight, note the difference and adjust what you read each time to what the shop scales read.

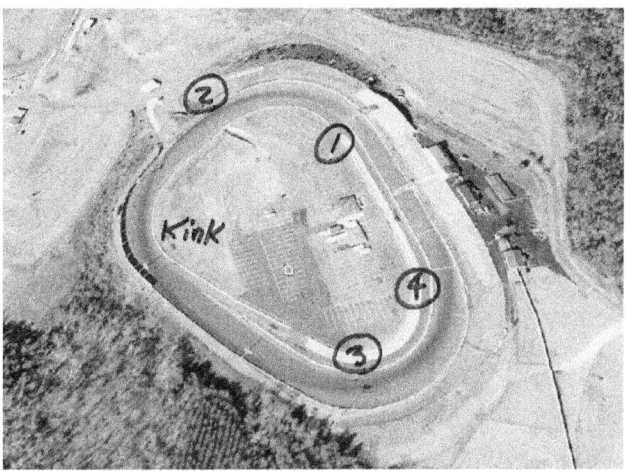

A well-prepared race team might print out an aerial photo of the race track they will be racing at. This view gives a lot of information about the turns and the line drivers have been using (the dark line around the track is where every one drives). This is Concord Speedway outside of Charlotte, NC. The configuration is different than most tracks. The part labeled "kink" is an extension of Turn Two because there is no real straight between Turn Two and the Kink. The length of the turning portions of this track is very long.

Find a spot at the track that is mostly level and park the car in the same spot every time it comes in. That way you can take measurements and set the sway bar the same each time without deviation in the numbers.

How to Measure Track Performance – We need to measure the on-track performance. There are two components to speed, the motor/gear, and the chassis setup combination. Since we work on these separately, we need to measure them separately.

If we have lap times that include turn segment times, we can then compare our times with our competition. Turn segment times tell us all we need to know about how good the chassis setup is verses the other cars we are racing against. Remember what we talked about, if we can improve the mid-turn speeds, we will usually also improve the straightaway speeds.

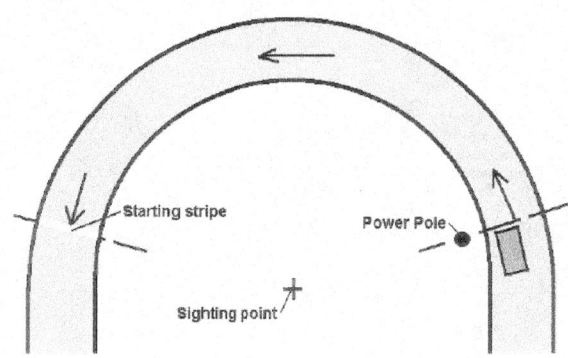

Be sure to take segment times that represent the speed through the turns. If you are several tenths off the fast whole lap times of your competitors, but just as fast through the turn segments, then the engine might need looking into. When you are the fastest car through the turn segments, you can stop making changes to your setup, or if you do, be prepared to go back right away if no further gain is seen.

The First Set of Runs - The driver should initially make several slower circuits and then a few faster laps to "shake down" the car the first time out. This establishes that the brakes work as expected, the wheels are on tight, the air will stay in the tires, there are no water or oil leaks and the transmission and rear end lubricants will be brought up to temperature. We should do two more longer runs following the initial outing before we can expect to get meaningful tire temperatures.

Have the driver initially run the turns at a speed lower than normal and note the position of the hands. Then once the car is up to speed, the driver should again note where the hands are and if the steering is significantly different, the car is either tight or loose.

After each run, record the tire pressures and temperatures, tire sizes, and engine water and oil temperatures. Keep hard copy records of the data in addition to digital records that may be stored in the tire temperature device or a computer.

Once the driver is confident that the car is sound, longer and faster runs can be done. As you make your next series of runs, try to have the driver stay out at least 10 laps so that the tire temperatures will be sufficient to show how they are working. The exception to this is when the car is obviously not right. Then the driver should come in and explain the problem.

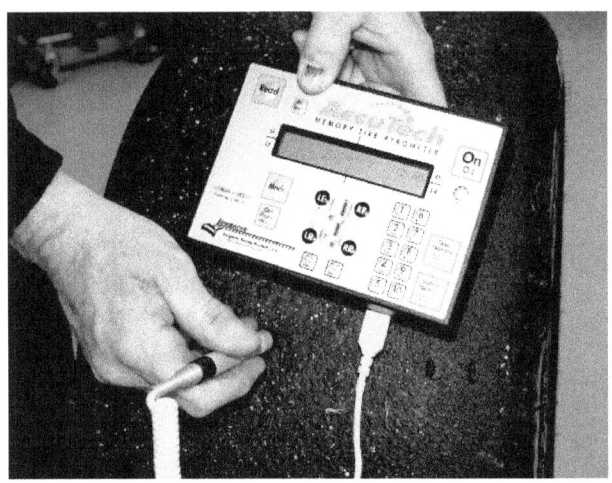

Record your tire temperatures and refer to them for tire pressure adjustments and camber changes. Bring the tires up to race temperatures and analyze your tire data. Correct any camber or tire pressure problems before making major setup changes. The priority starting out is handling balance (i.e. neutral handling), then tire temperatures related to pressures and cambers, then on to setup balance.

When we first go out and run hot laps, we need to quickly react to the tire temperature and make changes to the tire pressures and cambers before we make any major setup changes. The pressures and cambers help to create the largest tire contact patch possible. This adds a lot of grip to a tire without ever making a spring or other setup change.

Initial Evaluation - Evaluate the tire cambers, pressures and overall handling balance. Make quick adjustments to the front tire cambers and all four tire pressures if the temperatures dictate. Do not make chassis adjustments until the tire issues have been corrected.

Record the driver comments as well as crew comments as to the handling and engine performance. If the car is not neutral, now is the time to make changes to improve the handling while working to maintain a balanced setup.

There is a difference between Handling Balance and Dynamic Balance. The car is neutral in handling balance when it is neither tight nor loose. We can easily adjust most cars so that they will be neutral. This may make the car faster, but it is not our primary goal. We need the car to be both neutral in handling and balanced in how the front suspension and rear suspension are working.

The first runs should be made to shake down the car. We want to make sure there are no fluid leaks, and that the brakes and engine are working correctly. The driver can relay comments about the general handling once the car gets up to speed.

Evaluate the data from every run. If possible, try to use a data acquisition system to record shock travels, steering, throttle and more. Keep good notes as to your progress and how each setup change affected the car and the way the driver feels. Keeping detailed notes helps to refresh your memory so that you can look over the progress of the test later on in the quiet of your shop. Convert written notes to computer notes so you won't need to worry about losing them.

Mid-Turn Performance First — We need to always evaluate and correct the mid-turn performance first. To balance the car at this Steady State point on the track will also help to balance it on entry and exit.

We can interpret the balance of the car by evaluating the tire temperatures. These tell us how much work each tire is doing in relation to the other tires. We are looking for more equal temperatures on pairs of tires on each side of the car.

Changes to the panhard bar height and/or spring rates can help add temperature to a tire that is too cool. More uneven front tire temperatures indicate a tight car. High RR tire temps indicate a loose car. Once the tire temps come to be more equal on each side, the handling balance must be tuned with cross weight.

Remember that a tight car will also often be loose off the corner. This is called "tight-loose" because the initial condition, being tight, leads to the other condition, becoming loose off, and fixing the tight condition will often solve the loose condition.

If the car were tight with the original setup, to make it neutral in handling, the team might have found a cross weight percent that worked to make the car more neutral. As we make changes to the setup to heat up the LF tire, the car will then begin to have more front grip. This can make the car loose, so we then need to increase the cross weight to bring it to a more neutral handling condition.

Remember that spring split in the rear and panhard bar height changes are the most effective ways to re-balance a circle track car. As you make those changes, you will also need to tune the handling balance with changes in cross weight.

Tire wear can tell us a similar story when racing on dirt. Dirt teams rarely take tire temperatures. They do feel the tires for temperature, so we know that they consider the temperatures to be important. But tire wear can also tell us how hard a tire is working.

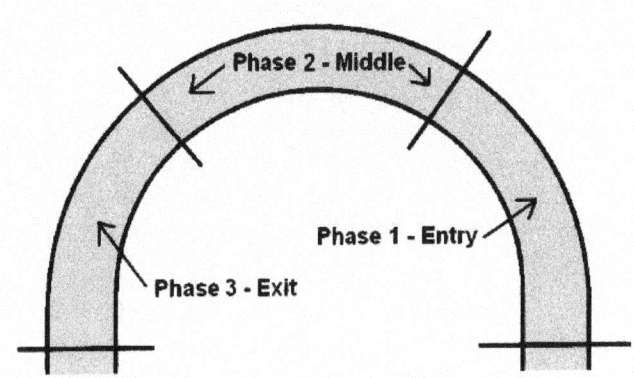

For tuning the setup, there are three turn segments that we work on. The place to start is in the mid-turn segment where poor handling and dynamic balance will affect the other two segments. We perfect the mid-turn portion first, and then the entry and exit portions in a way so as to not affect the mid-turn performance we have gained.

Mid-turn Changes - To change mid-turn balance, we can do one or more of the following:

1) Raise or lower the rear Moment Center by moving the panhard bar or J-bar up or down. For leaf spring cars, we can raise or lower the actual spring, but that is not an easy thing to do. Metric four-link cars also have a tough time changing the rear moment center height and must rely on other methods for changing the balance.

NOTE: For road racing cars with AA-arm front and rear suspensions, the roll centers can be changed to effect a change in handling. In the authors experience, changes to the roll centers in AA-arm suspensions have significantly improved the grip for that pair of tires. Lower roll centers help create more roll angle. Changes to the upper control arm angles can improve the camber change characteristics to help improve the size of the contact patch.

2) You can change the rear spring rates. Softening the right rear springs, and/or stiffening the left rear spring will increase the rear roll and will tighten the car, as will softening both rear springs. The inverse is true, stiffening the RR spring and/or softening the LR spring will loosen the car.

3) Softening the front springs will help the car turn, but to a lesser degree than making rear spring changes. Spring split at the front also has less affect and has more influence on entry characteristics than on mid-turn.

4) Installing larger or smaller sway bars will have an effect on handling. The stiffer the bar, the less the front

will want to turn. So, to help cure a tight car, we can go to a softer sway bar.

5) Increase or decrease the cross weight percent. As we make changes to the cross weight, we affect the handling of the car and we can easily make the car neutral in handling by making cross weight changes. But, this is not the ideal method by any means, it is just the easiest.

6) We can increase or decrease the stagger. This is never an acceptable way to tune the handling of your race car. For a defined radius of turn, there is an ideal stagger that will allow the car's rear wheels to roll around that radius and not influence the direction the car travels from following that radius. Learn what stagger your track needs and use that.

Summary – In this Lesson, we talked about our goals for testing, now to plan out a test and how to tune for mid-turn, the first consideration. In the next Part Two of this testing theme, we will tune for Entry and Exit performance and explain how to evaluate the results of your test.

Exam - In The Context Of This Lesson:

The Overriding Goal For The Test Is To?

1) Find the fastest lap times

2) Create a dynamically balanced setup

3) Tune the setup for a perfect handling balance

4) Compare our lap times to our competitors

The Order We Need To Solve Handling Problems Is?

1) Entry first, then middle and then exit

2) Exit first, then entry and then middle

3) Middle first, then exit and then entry

4) Middle first, then entry and then exit

Road Racing Cars Easily Cannot Change What?

1) Weight distribution front to rear

2) AA-arm Roll Centers

3) Cross weight percent

4) All of the above

The Very First Changes We Make Are?

1) To the spring rates

2) To the panhard bar heights

3) To the tire pressures and cambers

4) To the shocks

To Correct Dynamic Balance Issues, We Change The?

1) Spring rates

2) Panhard bar heights

3) Sway bar sizes

4) All of the above

To Correct Handing Balance Issues, We Change The?

1) Spring rates

2) Panhard bar heights

3) Sway bar sizes

4) Cross weight

Lesson Twenty-One – Tuning At The Track, Part Two

Entry Tuning – When evaluating and making changes to the car to correct entry problems, do not make a change that will affect the mid-turn portion of the track. You should have already tuned the mid-turn handling and balance. If you must change a spring to help entry, then you need to re-tune the mid-turn balance.

Entry problems can be caused by rear alignment issues or incorrect shock rates, mostly in the LR corners of the car. Over-driving the entry can make a car push. Make absolutely sure that the rear end is aligned properly and square to the centerline of the car. Do not install a high rebound left rear shock. And don't let your driver "dive bomb" the entry to the corners.

Excess LR shock rebound may cause the car to be loose on entry as load is transferring to the front while braking. The LR shock should allow the LR tire to move in rebound to help it maintain contact with the racing surface as the car pitches forward and to the right on entry.

Spring split has some effect on entry performance too. At flatter tracks, a stiffer LF spring over the RF spring helps entry stability in most cases. Remember that spring changes also affect the Dynamic Balance of the car and you will need to re-evaluate the tire temperatures and make changes to the panhard bar height to re-balance the setup after a spring change.

On entry, the car is turning in, braking, and transferring weight that is mostly going to the right front corner. We need to provide more compression control with the RF travel and less rebound control for the LR as weight comes off that corner.

Entry Changes - To change entry performance, we can do one or more of the following:

1) Rear alignment is the number one cause of entry problems. The cause relates to either the miss-alignment of the rear tires or by the rear steering of the rear end. A car can become tight or loose on entry and that can translate to mid-turn problems. You should have checked and corrected any rear alignment problems long before you came to the track. Rear alignment and rear steer are not mid-turn tuning tools.

2) Shocks affect entry. Shock rates that restrict movement of one or more corners of the car can negatively affect entry. A LR tie-down shock, or one with excessive rebound control will help cure a tight-in condition by loosening the rear, but this is considered a crutch and not a valid tuning tool.

The two corners most affected by the dynamics of corner entry are the LR corner and the RF corner. A RF shock that is too stiff on compression can cause a tight condition on entry and a LR shock that is too stiff in rebound can cause a loose condition on entry.

3) Brake bias changes affect corner entry. There is an ideal brake bias that will allow adequate braking of the front and rear sets of tires based on the loads those tires carry. Different cars with different Center of Gravity heights will require different brake bias.

Tune your brakes so that wheel lockup occurs simultaneously at the two ends of the car under heavy braking. We do not want the brake bias to influence entry handling characteristics. Never try to correct a car that is tight into the corner by increasing the rear brake bias or fix a loose-in car by increasing front brake bias.

4) Setup Changes to solve corner entry problems? We never want to make changes to our spring rates, sway bars, weight distribution or moment centers to solve entry problems without making sure we then bring back the mid-turn balance. If you do that, we will certainly change your mid-turn handling in a negative way. You should have already tuned the car so that the mid-turn handling was balanced correctly.

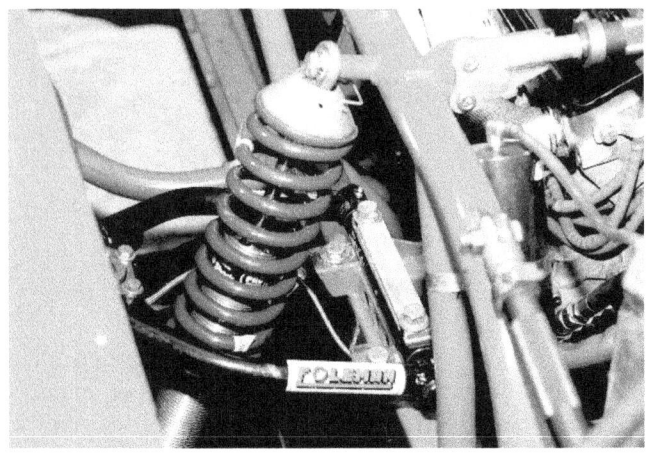

Be sure to match your shock rates to your spring rates. It's OK to go a little higher on the rebound rate verses your spring rate, but you don't want to install a 'tie-down", or very high rebound setting where the spring doesn't require that. Use a normally valved shock on a corner with a more conventional spring rate. High rebound shocks are designed to be used with bump stops and bump springs mostly, and certain soft spring conventional setups that we described in Lesson 17.

Exit Tuning – Problems associated with corner exit involve either a tight-off or loose-off condition. If we use the wrong methods to improve either of those, we might then end up with a car that no longer handles in the middle. So, the changes we make to improve exit performance should never change the mid-turn balance. Changes to spring rates, spring split, panhard bar height and cross weight will all affect, and probably ruin our mid-turn balance. So, just how do we tune exit performance?

The tracks where we usually see exit issues are mostly the flatter tracks with associated lack of grip or at tracks where the transition in the banking cause problems. The combination of lateral forces that come from turning the car and the torque associated with power application tend to overload the grip capability of the rear tires. So, we need to develop ways to increase the amount of grip the rear tires have available on exit while being careful not to affect the mid-turn balance we have established.

We can experiment with various designs of Pull Bars, Push Rods, Lift Arms, and rear steer that happens only on acceleration. The goals are to reduce the "shock" to the rear tires upon initial application of power and increase the total rear grip by introducing rear steer (to the left) into the rear geometry. The more the rear tires are steered, the more traction they will develop, just like the front tires when they are steered. This is the "angle of attack" syndrome we learned about earlier.

There is a limit to how much rear steer we can use before the car becomes too tight and we cannot steer the front wheels enough to overcome it. A few ten-thousandths of an inch of fore and aft rear wheel movement can be felt by the driver on asphalt. Most movements are in the range of 0.050" to 0.125".

One design element uses about 0.063" of RR wheel movement to the rear on entry, and then 0.125" RR wheel movement to the front on exit. This steers the car to the right by 0.063" on entry and a net of 0.063" to the left on exit.

Loose Off Condition – We can use Rear Steer to solve loose off problems. If we know we are good through the middle, then a loose off condition can be solved with the application of rear steer that happens only upon the application of power.

The most common way to create rear steer in cars equipped with the three link suspension is to increase the angle of the left trailing link, the front being higher than the rear. If we also decrease the anti-squat by decreasing the angle of the third link, when the car sits down on turn exit under acceleration, the LR wheel will be moved back creating rear steer.

We can tune the amount of rear steer by changing the angle of the LR link. This works very well and compensates for the rear tire drift caused by adding power. The rear tires will attain a new angle of attack to gain traction and if we cause rear steer, we immediately give the rear tires that angle of attack it needs without needing to step out.

Shock rates can temporarily increase the cross weight percent on exit to tighten your car off the corners. If you run shocks with a stiffer compression rating on the LR corner than on the RR corner, then when the shocks move as the car squats coming off the corner under acceleration and while the loads transfer to the rear, then the LR corner will momentarily carry more load and the LR and RF will then share that increased load.

Throttle Control - This is a learned art and will allow the rear tires to maintain their grip on the track surface and help to provide better acceleration. Once we lose grip in the rear, we must back off the throttle until we regain grip before we can continue to accelerate. By exercising throttle control, we may feel like we are giving up performance, but in reality, we are providing the most acceleration possible.

Throttle control is defined as the modulation of the gas pedal through a range of motion, never moving quickly from one position to another, in order to keep the tires in contact with the track surface. The rate of change in throttle position must be altered depending on your position on the track and through the corner, so the driver must develop an educated foot.

Many dirt drivers report that they never got past half throttle over the course of an entire race in which they won. This means that they were working from off throttle to half and many points in between. It is the development of efficient throttle modulation that is one of the most effective tools you can use to promote bite off the corners.

We can utilize rear steer to gain entry performance and to create more rear tire angle of attack. We have described ways to create rear steer and how much to use in previous Lessons. Make sure you test these configurations before trying to race with them. A little rear steer goes a long way.

Balance Evaluation Using Tire Temperatures – The question becomes, how do I know when the race car setup is balanced dynamically? There are several ways to evaluate the setup for dynamic balance. One of the first and most simple ways is to compare the average tire temperatures for each of the four tires. By now, we probably won't have any one tire that is way hotter or cooler than the others.

We mostly look at the left side tire temperatures for circle track cars, and front to rear average temperatures for road racing cars. In today's circle track world, the LF tire may be either cooler or hotter than the LR tire. Let's clarify that.

LF Cooler – If the LF tire is cooler than the LR tire, and the setup is more conventional in spring layout and having two positive roll angles, then the setup is tight. That is, the rear is rolling more than the front. This takes more load off of the LF than would otherwise transfer due to lateral load transfer. The front pair of tires end up with less grip than the rear pair of tires.

LF Hotter - If the LF tire is hotter than the LR tire, then the front is out-rolling the rear. This means that we have arranged the spring rates such that the rear wants to roll less than the front. This will transfer more load off the LR tire than would otherwise transfer due to lateral load transfer.

The situation with the LR being cooler than the LF may be due to a setup where the front and rear roll angles are equal and opposite, as we have discussed. The front wants to roll to a positive roll angle and the rear wants to roll to an equal value, but negative roll angle. These two will then cancel out each other, but this leaves the LR less loaded than if the two roll angles were the same and both positive.

Chart 1
Session #3

Left Front

Loc 12	Loc 11	Loc 10
146	154	150
Avereage Temp.	150.00	
Pr. Cold	Pr. Hot	Pr. Gain
20	23	3

Right Front

Loc 1	Loc 2	Loc 3
211	204	207
Avereage Temp.	207.33	
Pr. Cold	Pr. Hot	Pr. Gain
24	29	5

Front Average = 178.67

Left Rear

Loc 9	Loc 8	Loc 7
162	168	158
Average Temp.	162.67	
Pr. Cold	Pr. Hot	Pr. Gain
20	25	5

Right Rear

Loc 4	Loc 5	Loc 6
216	214	210
Average Temp.	213.33	
Pr. Cold	Pr. Hot	Pr. Gain
24	32	8

Rear Average = 188.00

Tire temperatures can tell a lot about how the chassis is working and if the setup is balanced or not. In this example, we see where the LR average tire temperatures are almost thirteen degrees hotter than the LF temperatures. And, the RR tire temp. is six degrees hotter than the RF temps. This car is setup tight based on the cooler LF tire temps. and is probably loose off the corner. This is commonly called, "tight/loose" because the tight condition through the mid-turn phase actually causes a loose off condition. This also can make the rear average temperatures hotter than the front temps.

Balance Evaluation Using Force Evaluation – Another way to determine the dynamic balance is to physically measure the right front force that the spring and sway bar are together putting on the RF tire. For every type of race car running a certain type of race track, there is a combination of the four tire forces that is ideal. This situation means that the perfect equal-unequal tire loading is taking place at the front and rear.

This has been covered in past Lessons, but to review, ideally, we need for the front and rear pairs of tires to be equally un-equally loaded. A pair of tires on any race car in todays racing will never be equally loaded when going through mid-turn. Because we have load transfer from the inside tires to the outside tires, the inside tires will always be less loaded than the outside tires.

We call this un-equal loading. We want the front and rear to be equal in the un-equal loading so that the outside pair of tires carry the same tire load, and the inside pair of tires carry the same tire load.

Under this situation, if we can measure what the RF tire load is at mid-turn, then we can adjust the setup so that the force supporting that tire provides the perfect tire loading. But, we have to know what force we are looking for.

In a typical super late model circle track car, the RF tire load at mid-turn might be around 1,500 pounds. If the spring motion ratio were say 0.8125, then the MR squared is 0.6602. If we divide that number into 1,500, we get 2,272 pounds of spring force needed to support that mid-turn tire load.

There are many different products that will measure the spring force if we know the exact shock travel at mid-turn. We support the Gale Force Load Pro because it was the first on the market and offers excellent tech support to the user.

Remember too that the force needed to support the RF tire load is often a combination of the ride spring, the bump device and the sway bar spring rates. Measuring the ride spring and bump spring rate combination is fairly simple. It takes a special tool to measure the sway bar spring force. Gale Force also has a product that will measure the sway bar force at the shock/spring position. This is important because we need to be able to add these two forces together.

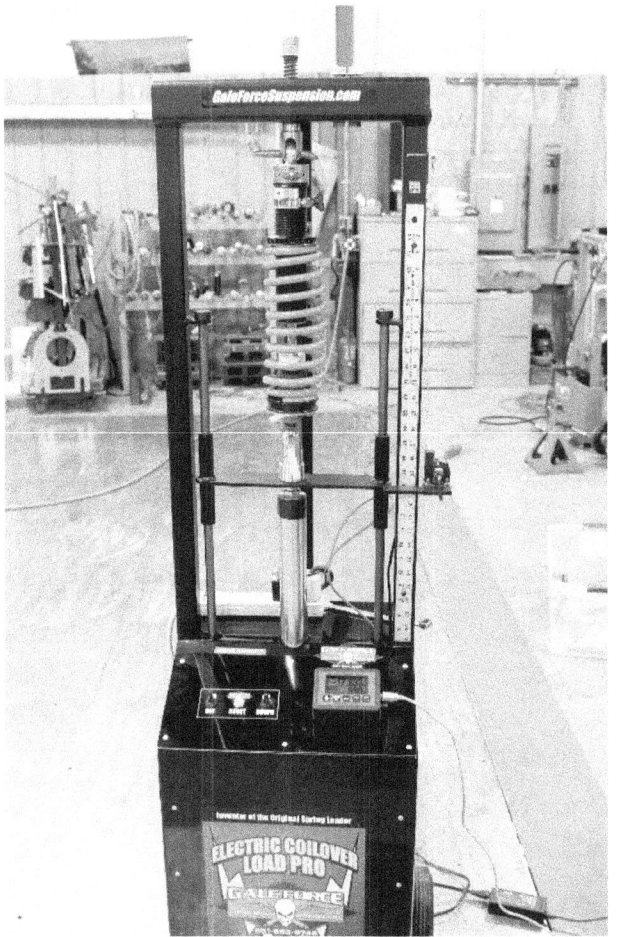

The Gale Force Load Pro rig allows a team to re-enact the forces that occur on the race track. If we know the shock/spring travel at mid-turn, then we can then move the coil-over to that same length on the Load Pro and then record the spring force. There is an ideal spring force for the RF that will provide the ideal tire loading for that car on a particular track.

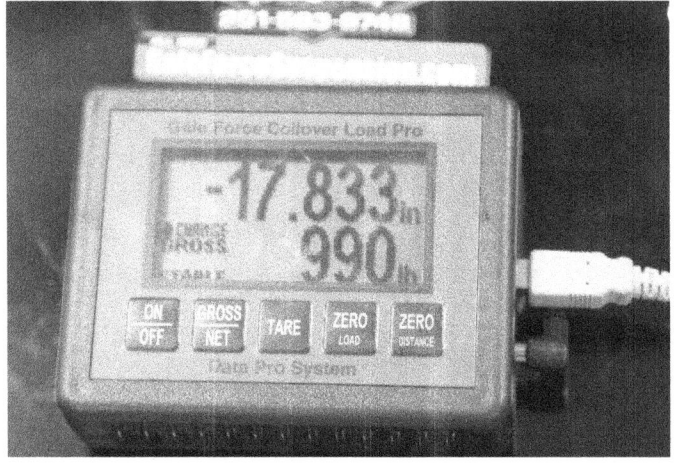

Here we see the readout from the Gale Force Load Pro. The shock length is 17.833 at ride height and the force reading for the ride spring is 990 pounds. When we compress the shock/spring and bump combination to the length it is at mid-turn, then we can

read the force that the spring combination provides to help hold the RF tire load. The sway bar is a part of the combined force that supports the tire load, and it can be added to the spring and bump forces.

The ride spring and bump provide much of the force needed to support the RF at mid-turn. The sway bar also contributes to that total force. Therefore, we need to know how much force is added by way of the sway bar. Gale Force also makes a sway bar tool that measures the sway bar force at the coil-over position so that the ride spring/bump force can be added to the sway bar force. This gives us the total spring force supporting the RF tire.

Once we make adjustments to the setup so that the RF spring force is correct, then the other tire forces must also be correct because each tire force is interdependent on the others. If one changes, then the other three also change. If 2,272 pounds of combined spring force is what we need, then we adjust the setup until we read the shock travel that will supply that force.

Errors In Shock Travel Readings – It is possible to read errors into the shock travels. If you are reading shock travels using a travel indicator mounted on the shock body, then you need to know that the travel represents only the mid-turn shock travel and not travel from any other influence.

Another influence might be due to braking into the corners. If the RF shock travel we see on braking exceeds the shock travel that happens at mid-turn, then our reading is in error. We might think there is more force on the tire than there actually is at mid-turn.

The way to avoid this is to either use data acquisition and read only the mid-turn shock travel, or run laps that avoid heavy braking into the corners, but provide maximum speed through the mid-turn portion. The driver needs to work with the team to accomplish this.

Another way to reduce the braking influence is to use Anti-dive geometry in the RF and LF upper control arms. This helps prevent over-travel of the shock while entry braking. The Anti-dive will not interfere with any other aspect of the setup if applied in measured amounts.

Normally a team will use a shock travel indicator such as the one seen here. This is adequate if the user knows how to eliminate errors in the reading. On some tracks, entry braking that causes the nose of the car to dive, can put more load and shock travel on the RF suspension than what we would see at mid-turn. So, care must be taken to run a set of laps where braking is diminished and the maximum shock travel is due to mid-turn suspension travel.

Here is an example of entry braking where the RF has traveled probably more than what it will at mid-turn. Anti-dive geometry can help solve this problem that will cause errors in our shock travel data.

One of the solutions to errors in shock travel readings is to use data acquisition on the car during testing. Then the team can look at the data and isolate the shock travels through the mid-turn portion. Any additional travel that might come from entry braking, or even transitioning up onto the banking when going out for a run, can be isolated.

End Of The Test - Always save your sticker tires for the last run of the day after the car is all dialed in. If the setup is good, make a qualifying run on fresh tires. After that run, do a 30 or 40 lap run on those newer tires and see if the lap times stay consistent. A truly balanced setup will provide lap times that fall off less than your competition as more and more laps are run on a single set of tires.

Review your notes when you get back to the shop and learn from both the gains and losses. All of the results are valuable and the more we learn about the effects of changes, the better we can make quick adjustments during a racing event. The top teams make a point of knowing how each chassis adjustment affects all of the other parameters involved with their setups.

Incorrect tire stagger, bent shocks, suspension binding and poor alignment are some of the peculiarities that can ruin a test session. If radical setup changes do not seem to provide the expected result, look for a mechanical problem and fix it. Keep your test notes available for review. Test as often as you can afford and whenever the track is available. If you can develop a comprehensive plan for your testing, your performance will get better and you will enjoy your racing experience that much more.

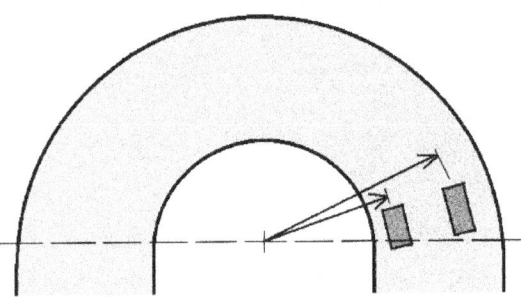

Radius to inside tire = 200 feet or 2400 inches (12" x 200')
Radius to outside tire = 2465 inches (2400 + 65" track width)
Outside tire travels 2465 x 2 x PI (3.1416) ÷ 2 = 7744 inches
Inside tire travels 2400 x 2 x PI ÷ 2 = 7540 inches
Outside tire = 85 inches in circumference
Inside Tire = 85 x (7540 ÷ 7744) = 82.75
Correct Stagger = 85 minus 82.75 or 2 1/4 inches.

The correct stagger you will need for a particular track is fixed and not to be used as a tuning tool. We never want to adjust stagger to mask a handling problem. Using incorrect stagger is a crutch and will only provide a temporary solution. Running the wrong stagger can cause a multitude of handling problems. The example shows a typical calculation for stagger on a track with a 200 foot radius turn.

The data showing the corner the tire was mounted on, the track it was run on, the date and the set numbers are important things to keep track of. Some teams will write that information on the tire. Keeping the information printed on the tire saves time and confusion when deciding which tires you need to put on the car next.

Conclusion - The test procedures are based on the assumption that you have already solved the most critical issues facing your race car. Back at the shop, you have aligned it, checked and corrected the moment center design, checked for binding in the suspension, rebuilt the shocks, set the weight distribution and done all of the other maintenance things we know we should do.

The last thing to do is run the car. Teams that have learned how the various changes will affect the handling of the car will go through the process much more quickly. If you're just now learning these things, take good notes and concentrate on what is happening with each change. Ask your driver a lot of questions so that you will know exactly what changes to make and how far to go with each change. Once you complete the test and have fine tuned the setup, it will be time to go racing.

Exam - In The Context Of This Lesson:

Loose-In Problems Might Be Caused By?
1) Brake balance issues
2) Shock valving that is too high in rebound
3) Rear steer issues
4) All of the above

Tight-In Problems Might Be Caused By?
1) Brake balance issues
2) Too much right front shock compression
3) Overdriving the corner
4) All of the above

Loose-Off Problems Might Be Caused By?
1) Rear steer issues
2) A stiff RR spring
3) A car that is tight in the middle
4) All of the above

Tight-Off Problems Might Be Caused By?
1) Rear steer issues
2) A soft RR spring
3) LF shock rebound too high
4) All of the above

Tire Temperatures Tell Us What?
1) If our spring rates are too high
2) If our spring rates are too low
3) How the tires are working
4) When the weight distribution is wrong

Reading The Right Front Total Spring Force Tells Us What?
1) How the tire temperatures will look
2) The loading on the RF tire
3) If the RF spring rate is too high
4) If the RF spring rate is too low

The Sway Bar Does What?
1) Resists chassis roll
2) Adds RF spring rate
3) Adds to the RF suspension force
4) All of the above

Errors In Reading Shock Travel Come From?
1) Misreading the tape measure
2) Going up onto the banking
3) Chassis dive from braking into the corners
4) All of the above

Race Car Technology – Level Three
Lesson Twenty-Two – Post Race Analysis

We now have one race under our belt with the changes we have made. The true measure of success, or of making the right decisions, is how the race car performs in an actual race. The proof is in the pudding so they say and that might be one of the primary attractions people have to going racing and being a part of a race team. The results of your efforts are immediately known, to everyone.

That part, the knowing, can be a scary process to go through. Every "setup guy" knows exactly what I'm talking about. But, remember this one thing, every bit of information is valuable. Even the so called "failures" are not really that. They are, for the smart racer, opportunities to grow and learn. I've had my share of those.

It may take several races and an accumulation of information before your car is where you want it to be setup wise. And then there is the part that is the hardest to work on, the driver. In most cases, unless you have a proven, winning, driver, there is a learning process that every driver has to go through. Even for experienced drivers, the setup that is balanced and tuned correctly might feel different than anything they have experienced before. Just driving the balanced setup might take time to adjust to.

When you go through the process and work out a successful setup, there's nothing like it. To see all of your hard work end up in victory lane is priceless. We created ORS to help put racers on the right path to success. We can't do it for you, but with the right information and the right tools, you can surely do it yourself.

And, if the car was close, but not all the way there, the driver can have a better opportunity during the race to fine tune their thoughts on how to make it better, if not for anything else, his benefit.

Timing During The Race – The first point I will make about Post Race Evaluation is this. Make sure during the race that you have someone taking and recording lap times for your car as well as the leaders. This tells us a lot about our possibilities.

If your car is leading, then the lap times will tell us how much performance gains we have over our competition. If we are chasing the leaders, the lap times will tell us how much work we have to do to be competitive enough to lead and possibly win races.

If your car is off in lap times to the leaders, then it becomes necessary to know where the loss is being produced. You will need to have someone time your turn segment times against those of the leaders. This tells us a lot.

In a personal experience several years ago, I setup a Pro Late model car and felt we had a good balanced setup. In the first race we ran, the dad of the driver went to turns one and two and timed our car and the leaders. I went to turns three and four and did the same. He ended up telling me that we were about a tenth faster in one and two and I recorded laps that were the same times as the leaders at my end of the track.

We were off in total lap times by 0.5 second. The car was losing at least 0.25 seconds on each straightaway. This told us that we needed to work on the motor. If we didn't take the segment times, we could have easily assumed we needed to work on the setup, just like many teams would do.

If you are off on segment times, then obviously you need to find a setup that works to produce a faster mid-turn speed. Or, the driver needs to work on their driving style. This is a tough one to call. We'll be providing a driving school at some point in time and we'll address this issue in more detail then. But suffice it to say, in our experience, throttle and brake modulation is the difference between a good driver and a not so good driver. This has been proven over and over going back many years in racing. A lot of time can be lost between the lifting of the throttle and the point of mid-turn, due to the way the driver operates their foot.

That said, let's look at some things that might not have been perfect for your first race. Much of this is a review of previous information we have presented, but in the context of after the race evaluation.

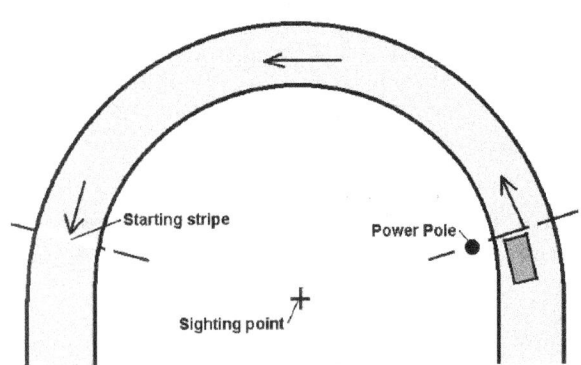

A major step in understanding where your car is performing, or not, is to take segment times for that part of the race track that is from entry lifting and the beginning of braking to where the car starts to straighten out onto the straights. The timing in this portion tells us how well our setup is working against the competition. It is a much better measure than total lap times.

Entry To The Corner – If the car had entry problems, think about when this problem occurred. Was it at the very start and got better as the race went on? If so, the drivers driving style might be something we can look at. Over-driving the corner on entry can cause a push from over-braking. This comes from the excitement of, well, the start of the race. If this might be the case, don't work in the car, work on the driver.

If the entry problem showed up later in the run, then maybe the RF tire is being used too much. This can be caused by too much front brake bias or maybe a tight setup. We can still have a tight setup while being neutral in handling. If other adjustments crutched the tight setup, then the tight part will eventually show up. Here is why.

When the race pace slows later on in the race, the actual lateral G-forces become less. When the G-forces go down, typically the car goes loose. This is a little advanced in our thinking process, but proven to be true. A good balanced setup will in most cases be balanced through a wider range of G-forces than one that is not balanced. I personally worked out a setup for a NE Tour modified team this year where the roll angles stayed the same through G-forces from 2.0 to 2.5 G's. The actual roll angle number went up as the G's increased, but the setup remained balanced. That is a wide range, but fairly typical of a truly balanced setup.

If the car on entry was good for one line, but not so good on another line, like good low but not so good high, then the shock package might need to be examined. There are usually different transitions in the change of the banking on entry and exit for different lines. The low line might have less abrupt transition than the high line does. Shocks can help the car transition through these changes in banking angle.

Entry tuning is the second phase of the on-track setup routine. When we get the middle worked out to create a balanced setup, many of the entry and exit problems will be solved. If your driver goes out and runs laps and has entry or exit problems, look first at the mid-turn performance before starting to change the setup to solve the other two segment problems.

Tight Or Loose Middle – The most obvious reasons for being tight or loose through the middle segment of the turn involves the basic spring/rear roll center parts of the setup, but then again, we were supposed to have worked that out in testing, right? If you didn't have a chance to test and just had short practice sessions before the race, then maybe when you put on those new tires at the start of practice, they masked the problem.

Definitely take tire temperatures immediately after the race if at all possible. These tire temps. tell more about the setup than the ones you collected from shorter practice runs.

If the tight or loose condition was present the entire race, then work on creating a more balanced setup by making setup changes. These longer runs will definitely provide more information than the shorter practice runs.

Look at the tire wear to determine if you are overusing one or more tires. If the car is setup tight, then the RF and RR tires might be abused. The RR tire will show more heat and wear than any of the others from either a loose setup, or a tight setup that produces the "tight/loose" syndrome, which we have talked about.

An Important Point - In yesteryear, the primary cornering problem usually involved a tight car because traditionally, most race cars didn't turn very well. What we know now about front geometry was not so widely known then. Now, in today's racing, we have learned how to produce larger contact patches for the front tires, and more than anything else, we have learned how to put a lot of loading onto the LF tire by using the bump setups and other methods.

Here is where the problem comes in. And this information comes as a result of a more recent experience that the owners of this school came across. With a dynamically loose setup, or one where the front has more grip than the rear, the rear tires will slip out and find an Angle Of Attack (remember that Lesson?) that will provide the grip they need to get through the turns.

This added AOA stresses the rear tires, especially the RR tire, and they cannot continue to provide this level of grip for a long time. In a race where we discovered the setup was very loose, the car ran fairly neutral for up to 30 laps before it began to go to handling loose. It never had great drive off the corners, but that could have been a natural result of the race track we were running on.

Only later did we discover that this car had a much lower CG than originally estimated. So, the setup we installed was loose for that reason. We learned something from this. A dynamically tight setup becomes evident much sooner than a dynamically loose setup.

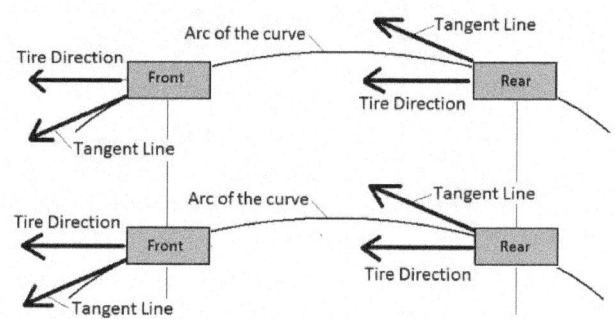

We have learned that a car will be more forgiving when the setup is loose verses when it is tight. Some of that is dependent on the driver and what they like. Some drivers can drive a loose car, some cannot. For those who can, a setup that is more towards the loose side can be faster. That is because as the rear slips out, it finds more Angle of Attack and gains grip. The car then becomes more neutral and not so loose.

This photo is taken near mid-turn and we can see where the rear tires have slipped out and are now at a higher angle of attack to the direction the car is traveling. This is exactly what we are talking about in this section. By running at this angle, the rear tires have more grip and the loose condition goes away.

Exit Performance – If your car is good on entry, good through the middle, but loose or tight off the corner, again, when did this happen? In the previous case of the Dynamically Loose setup we described, as the car began to go loose through mid-turn after the 30 lap point, it also became very loose off the corners.

So, in contrast to the tight/loose setup that is very common throughout the past twenty years or so, we now have a Loose/Loose syndrome happening. The car is setup loose but doesn't show it until later in the run.

If in a race, because of many caution periods, the car happens to run short sprints, the loose condition from the loose setup may never show up. But, the drive off the corner will never be as good as it could be. That might just be a clue and something to consider when evaluating the actual race performance.

Going back to the case of a tight/loose setup, the loose off condition will show up fairly early in the run, or race, because the rear tires don't have as much time to seek the added AOA that will help mediate the problem. Just as the rear tires are slipping out, the driver gets into the throttle and the rear tires cannot adapt that quickly. The driver will report a loose car.

When a race car is exiting the turns, there is a lot to consider that is different than what we do for entry and mid-turn. We still don't want to disturb the mid-turn performance by making wholesale spring and roll center changes that will disturb the mid-turn performance.

How To Tell If The Car Is Balanced - The key to realizing and recognizing a truly balanced setup is how the car performs over a longer number of laps run. In one of my first experiences with a race team, we had developed a truly balanced setup. We ran a full 50 laps on sticker tires at a test before the season started. The lap times fell of about 2.0-2.5 tenths of a second by lap 20 in the run. The next 30 laps, the times did not fall of at all. They stayed very consistent within a few thousandths per lap. That car won a lot of races.

The goals for a team are up to the team to determine. If the races are short, 35 laps or less, and there will inevitably be a few cautions thrown in, then maybe a jack rabbit setup that won't be good in the long run might just be the best setup to run. For those conditions at that kind of track, maybe this kind of setup will work better. Only testing and racing will tell.

If your racing involves longer races beyond 35 laps, and in the later laps there are usually less cautions to break up the runs, then by all means, the balanced and more consistent setups will usually beat all others.

It is true of most racers, even in today's racing, that they test and practice to find a setup that will produce the fastest lap times. These setups will often be un-balanced in either the tight or loose direction, depending on the type of race track. The team can definitely get away with this if the runs and the race duration are shorter. If your type of racing involves longer duration and longer runs, then the totally balanced setups will be superior in the long run, literally.

One of the benefits of a balanced setup is the ability for the driver to run different lines, high or low, through the turns. In a test we did, a modified Tour car was balanced through a range of 2.0 to 2.5 G's. In many cases, as the tires wear, a group of cars will slow from a half second to a full second or more in the course of a race. Since G-forces are dependent on speed through the turns, as the tires wear, the turn speeds decrease and the G-forces go down, So, a more consistent setup that stays balanced through a wide range of G-forces will stay balanced the entire race.

There will be situations where you will be forced to run different lines, so a balanced setup is critical in being able to compete and be fast no matter what groove you end up in.

Summary Lesson Twenty-Two – The actual race event is where we can learn a lot about the way our car is setup. The setup is only part of the equation, the driving being the other part. We can evaluate the tire temperatures and tire pressures and we can determine that we have a good and dynamically balanced setup. Once we know that, we can move on to evaluating our motor and gearing combination, and also to driver considerations.

Gather the data and record the data. A lot of time is spent looking over all of the information gained in testing and the race by experienced and competent crew chiefs. Make information collection a priority.

Most of all, use the data to make a decision and then go with it. Don't go back and forth with your thoughts. Too many teams do this. The race will tell you what you need to do next. Trust what the race tells you and make adjustments for the next race. It's a process that is fun if you do it right, because success is its own reward. And once you find that perfect setup, don't change a thing, just maintain it.

Exam - In The Context Of This Lesson:

Which Information Is The Most Valuable?
1) What works to make the car better
2) What works to make the car worse
3) Neither, both are valuable

Lap Times Tell Us Where Our Performance Is Lacking?
1) True
2) False

Entry Problems Can Be Caused By Which?
1) Brake bias errors
2) Rear alignment
3) Overdriving the corners
4) All of the above

Which Of These Might Show Up Earlier In The Race?
1) A tight setup
2) A loose setup

A Balanced Setup Will Be Good Where?
1) Through a wide range of G-forces
2) On the longer runs
3) Through the mid-turn portion of the turns
4) All of the above

How Can We Tell If The Setup Is Balanced?
1) By examining tire wear and tire temperatures
2) Less loss of lap times and performance
3) Consistent performance
4) All of the above

Race Car Technology – Level Three
Lesson Twenty-Three – Level Three Summation

Congratulations! You have just completed the Race Car Technology Level Three course. We hope you have gained a better understanding of your race car and what it takes to set it up. Whether you are coming from Level One all of the way to this finishing point, or just the Level Three, we have made every effort to provide the most current and useful information that exists in today's racing community.

For those who have taken the complete journey from Level One, you have learned about all of the parts and pieces that make up a modern-day race car. Then when you worked through Level Two, you learned how to work with those parts to assemble and make correct all of the chassis components in preparation to setup the race car for competition.

We taught the students, in Level Two, how to adjust all of those parts and pieces so that they would work together to improve the performance of the race car. We discussed bump steer, alignment, Ackermann, rear steer and much more.

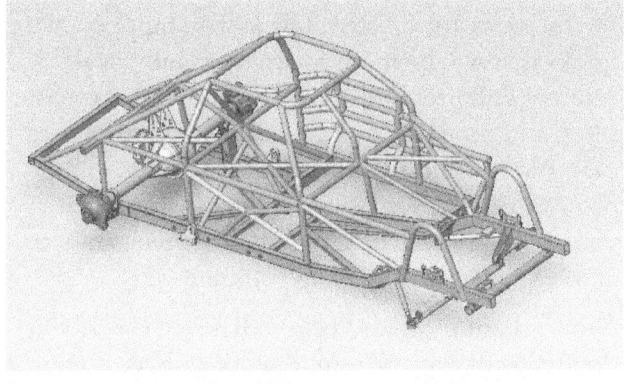

We started out with our Level One courses that taught the students all about how a race car was constructed and named all of the parts and pieces that go into a modern day race car.

Now in Level Three, we learned about the engineering that goes into the decisions we will make about how to setup a race car. We now know important information about both the car and the race track we will be racing on that will make us successful.

This all helps us to make intelligent decisions about both the initial setup we will use, but also how to tune that setup once we hit the race track. We now know how to evaluate the setup during testing or practice and what to do after we have raced the car in actual competition.

The racing environment is ever changing and the way we design and setup the race cars is also changing as time goes on. Just when you think you've got it all figured out, someone, or some group comes up with another plan and takes all of this to a new level.

The level we are at right now, in April of 2018 as I write this Summation, is very advanced from when I started my career some twenty years ago. But, that being said, the basics are still the same as to how to best make the four tires work. IN todays racing, we are now causing the cars to run at a much different, and more advantageous, attitude than ever before.

This thing called Race Car Setup is not magic. It is science and the overall theme is one that everyone can understand because it is simple in concept and fairly easy to apply. You just have to have the keys to the puzzle, and that in just a few words is what this school is all about. We provide the necessary keys to the puzzle.

Now, in this Level Three, we learned how to work with the forces and mechanical motions of the race car so that we can make our four tires gain the most grip possible for a much longer period of time. Understanding the forces and how to design the car to take advantage of those forces is what advanced race car setup is all about.

We started out with the intent of giving you the student the knowledge and the tools to be able to understand these basic principles and to teach you how to apply them to any race car. We hope we have done our job.

There will be no more school quizzes. The final exam is when you get to work on a real race car, hands on, and experience the thrill of going through the process of preparing that car for competition and then seeing it compete. It is both scary and exciting at the same time.

But, once you have learned the car and worked out all of the bugs, the first time you stand in the winner's circle, you'll look back with pride that you had the courage, fortitude and commitment to see this journey through to the end. That's the way it has been for the founders and all of our instructors in the Online Racing School. And I can tell you from personal experience, there's nothing like it.

You too can be a member of a race team, if you're not already involved with one, you just have to ask. Now that you have knowledge of how a race car works, and how to set one up, any team would be happy to have you join them and to be part of their success.

Thank you so much for trusting us to provide your advanced education in Race Car Technology. As to the future of RCT, we will be adding more course books that will be more focused on individual routines and processes involved in the preparation, setup and process of racing different types of race cars. We will enlist the help of industry professionals to act as the Instructors for those courses. These will be names that are well known and respected in our racing community.

Now, go out there and play with a race car. If you don't already have a team to work with, go to the race track and choose one. That is what I did and it worked out very well. Most teams are looking for educated and motivated team members. The fact that you have completed this school and showed an interest proves you are worthy.

Our publishing company will continue to grow and we will be adding more and more books as time goes on. Please keep checking back to look for additions to the school, as well as new content in those courses. Plus, we will be adding courses that deal specifically with routines and processes involved in setting up and racing in competition. These new courses will be presented by authors that are very experienced and well respected names in the industry.

Good luck in all of your racing endeavors and always stay safe.

Signed, Bob Bolles, Race Car Technology Courses

Exam - In The Context Of This Lesson:

Which Information Is The Most Valuable?

1) What works to make the car better

2) What works to make the car worse

3) Neither, both are valuable

About the Author

Bob Bolles has been a hands-on motorsports engineer for over twenty-five years. Although he holds a B.S degree in the Mechanical Engineering, his skill and experience with racecars comes from working directly on the chassis and with many hundreds of race teams. He likes to think that he comes from much the same mold as a few others before him such as his friend, the late Dr. Smokey Yunick, who also wasn't afraid to "get his hands dirty" to effect change. Like Smokey, Bob loves this sport and enjoys the interaction with others who also love it.

The nice thing about racing is that you cannot BS what you know or what you develop. It either works or it doesn't. Proof is just a few laps away. Before Bob started his research into racecar dynamics and engineering, he struggled with the very same problems that most race teams continue to struggle with today. He just knew there was a better way and if he just looked hard enough, the answers might come. Well, they did. That information is shared in the pages of the RCT series.

Over the years, Bob has worked with a high degree of success on virtually every type of stock car raced in the United States, as well as sophisticated formula type race cars. Whether it is an asphalt stock car, modified, or a dirt late model, his techniques and methods have improved handling and opened the door for winning. He has engineered cars that have won major asphalt late model championships, touring championship, modified championships, road racing championships, and dirt late model races including the Dream and the World 100 at Eldora Speedway.

It is this broad level of experience and his development of new technologies that qualifies him to write about these important subjects. Not a day goes by that Bob isn't speaking with or helping racers. With contacts all across America, he is truly in tune with the pulse of auto racing on a technological level that very few individuals enjoy.

The software Bob developed in the mid-1990's is still being used by championship winning teams throughout the U.S., Canada, Australia, New Zealand and Europe. His dream of being able to help all racers in their pursuit of success and enjoyment is fast becoming a reality.

Bob has been a professional motorsports technical writer serving as the Senior Technical Editor for Circle Track magazine for over fifteen years and currently as a technical contributor to Speedway Illustrated magazine.

Race Car Technology- Level Three represents Bob's third book in this series. Bob is also the author of RCT - Level One and RCT – Level Two. These books are being used by college instructors to educate future motorsports engineers.